Climate Change and Conflict in the Pacific

Shibata, Carroll and Boege address the various dimensions of the climate change–conflict nexus and shed light on the overwhelming challenges of climate change in the Pacific Islands region.

This book highlights the multidimensionality of the problems: political, technical, material, and emotional and psychological. Written by experts in the field, the chapters highlight the centrality and importance of opening up a dialogue between researchers involved in the large-scale global modelling of climate change and the local actors. Both scholars and civil society actors come together in sharing about the complexities of local contexts and the conflict-driving potential of climate change adaptation and mitigation strategies on the ground. The book brings together indigenous Pacific approaches with broader international debates in the climate change–security discourse. Through various accounts and perspectives, current gaps in knowledge are bridged, contributing to the development of more grounded, conflict-sensitive climate change policies, strategies, governance and adaptation measures in the Pacific region.

An important resource for students, researchers, policymakers and civil society actors interested in the multi-faceted issues of climate change in the Pacific.

Ria Shibata is a Research Fellow at the Toda Peace Institute in Japan. Her research explores the nexus between climate change, loss of land and identity amongst youth in the Pacific diaspora communities in New Zealand, Australia and Fiji.

Seforosa Carroll is a Lecturer in Cross Cultural ministry and theology at the United Theological College in Australia. She previously held the position of Programme Executive for Mission, specifically Mission from the Margins with the World Council of Churches based in Geneva, Switzerland.

Volker Boege is a Senior Research Fellow for Climate Change and Conflict at Toda Peace Institute. Dr. Boege has worked extensively in the areas of peacebuilding and resilience in the Pacific region. He is an Honorary Research Fellow, School of Political Science and International Studies at The University of Queensland.

Routledge Studies on the Asia-Pacific Region

Local Political Participation in Japan
A Case Study of Oita
Dani Daigle Kida

The US-Japan Security Community
Theoretical Understanding of Transpacific Relationships
Hidekazu Sakai

Opportunities and Challenges for the Greater Mekong Subregion
Building a Shared Vision of Our River
Edited by Charles Samuel Johnston and Xin Chen

Diversity and Inclusion in Japan
Issues in Business and Higher Education
Edited by Lailani Alcantara and Yoshiki Shinohara

Pandemic, States and Societies in the Asia-Pacific, 2020-2021
Responding to COVID
Edited by Charles Hawksley and Nichole Georgeou

For more information about this series, please visit: https://www.routledge.com/
Routledge-Studies-on-the-Asia-Pacific-Region/book-series/RSAPR

Climate Change and Conflict in the Pacific

Challenges and Responses

Edited by
**Ria Shibata, Seforosa Carroll and
Volker Boege**

Routledge
Taylor & Francis Group

LONDON AND NEW YORK

First published 2024
by Routledge
4 Park Square, Milton Park, Abingdon, Oxon OX14 4RN

and by Routledge
605 Third Avenue, New York, NY 10158

Routledge is an imprint of the Taylor & Francis Group, an informa business

British Library Cataloguing-in-Publication Data
A catalogue record for this book is available from the British Library

ISBN: 9780367431853 (hbk)
ISBN: 9781032593432 (pbk)
ISBN: 9781003001744 (ebk)

DOI: 10.4324/9781003001744

Typeset in ITC Galliard Pro
by codeMantra

Contents

Figures and tables

Figures

Table

Acknowledgements

We would like to express our gratitude to the Toda Peace Institute for providing the financial and logistical support which enabled us to organise annual workshops to bring together researchers, practitioners and policymakers from the Pacific Island countries, and around the world to engage in intense dialogue on the challenges posed by climate change and the potential pathways to conflict-sensitive mitigation and adaptation. This book is the culmination of our past conversations.

All of the workshop participants and authors in this volume deserve special praise for their willingness to offer their time, wisdom and expertise to explore the nexus between climate and conflict in the Pacific region. We would like to express our special thanks to Rosemary McBryde for her assistance in copy-editing the manuscript, and Chelsea Low for her patience in guiding us through the stages of the production process.

We hope that this book will enable its readers to understand the complexity of the climate change-conflict nexus in the Pacific and the overwhelming risks and challenges for the people of the region. It is our hope that this book will contribute to the development of increasingly conflict-sensitive climate change strategies in the Pacific region and a deep recognition of indigenous Pacific wisdom in relation to utilising local capacities for adaptation and sustainable development in the face of climate change.

Contributors

Volker Boege is Toda Peace Institute's Senior Research Fellow for Climate Change and Conflict. Dr. Boege has worked extensively in the areas of peacebuilding and resilience in the Pacific region. He is also Honorary Research Fellow, School of Political Science and International Studies, The University of Queensland. He works on post-conflict peacebuilding, hybrid political orders and state formation, non-Western approaches to conflict transformation, environmental degradation and conflict, with a regional focus on Oceania.

Halvard Buhaug is Research Professor at the Peace Research Institute Oslo (PRIO); Professor of Political Science at the Norwegian University of Science and Technology (NTNU); and Associate Editor of *Journal of Peace Research*. He leads and has directed a number of research projects on security dimensions of climate change, funded by the European Research Council, the US Department of Defense Minerva Program, the UN Office for the Coordination of Humanitarian Affairs, the World Bank and the Research Council of Norway. Recent publications have appeared in, inter alia, *Global Environmental Politics, Journal of Politics, Nature, PNAS* and *World Development*.

John R. Campbell has been researching population and environment issues in Pacific Island countries since the 1970s. He is currently working on the human dimensions of climate change adaptation and disaster risk reduction including environmental migration. He obtained a PhD at the University of Hawaii where his thesis was on population and environment interrelations on a small island in northern Vanuatu. He has written a book on development and disasters in Fiji, co-authored a book on Climate Change in Pacific Islands and a number of book chapters and articles on disasters, environmental management and global change, especially in Oceania.

Seforosa Carroll is an Australian Fiji-born Rotuman theologian who spent her formative years growing up in Lautoka, the Western side of Viti Levu in Fiji. Sef graduated in 2015 with a PhD in theology at Charles Sturt University. She is a CTI Fellow, was a resident member of the 2017–2018 Inquiry in Religion and Migration at the Center of Theological Inquiry, Princeton,

USA and is a Research Fellow of the Public and Contextual Theology Research Center (PaCT), Charles Sturt University, Australia. She was also a visiting lecturer in contextual theologies and interfaith dialogue with the School of Theology, United Theological College and Charles Sturt University. She was the Programme Executive for Mission, specifically Mission from the Margins with the World Council of Churches based in Geneva, Switzerland. She is currently a lecturer in Cross Cultural ministry and theology at the United Theological College in Australia.

Kirsten Davies' expertise is in international environmental law and sustainable development. She holds a master's degree (USyd), a PhD (USyd) in Sustainable Management and a PhD in Environmental Law (MQU). At the centre of her research is the engagement of communities and their connections with nature particularly in the context of climate change adaptation. Kirsten is the architect of Intergenerational Democracy, a method of whole-of-community engagement and capacity building and the subject of her (2012) book, *Intergenerational Democracy, Rethinking Sustainable Development*. She was awarded a Winston Churchill Fellowship to conduct research in the USA, the UK and Japan (2002) and was the recipient of the University of Sydney Agri-Management Scholarship for post graduate research (2005). Kirsten was awarded an Australian Government, Endeavour Award – Research Fellowship to undertake sustainability research in Vanuatu (2009) and maintains a close relationship with the South Pacific Region.

Kate Higgins worked as the Pacific Projects Manager at Conciliation Resources. Kate has been engaged in the Pacific region for over a decade in areas of community development, governance and peacebuilding. Kate has a PhD from The University of Queensland. Her research and practice interests are focused on the interactions between local- and state-based governance in the Melanesian region and how these interactions produce peace and livelihoods.

Matt McDonald is an Associate Professor of International Relations in the School of Political Science and International Studies at UQ. His research is in the area of critical approaches to security, and in particular the relationship between security and environmental change. He has published on these themes in journals such as *European Journal of International Relations, Security Dialogue, International Theory, International Political Sociology, Review of International Studies* and *Journal of Global Security Studies*. He is the author of *Ecological Security: Climate Change and the Construction of Security* (Cambridge UP, 2021) and *Security, the Environment and Emancipation* (Routledge, 2012), and is co-author of *Ethics and Global Security* (Routledge, 2014).

Ursula Rakova is an environmentalist and climate change activist from Papua New Guinea. In 2008, she received the Pride of PNG Award for her outstanding environmental contributions to the development of her country. She is the Executive Director of Tulele Peisa, a non-profit community organisation in Papua New Guinea and is responsible for organising the relocation of the inhabitants of the Carteret Islands to the mainland of Bougainville Province. Ursula is also a strong advocate of human rights, and has set up community schooling for young Bougainvilleans, and is a campaigner for the survival of her people.

Ria Shibata is a Research Fellow at the Toda Peace Institute, Japan. Her research focuses on the role of challenged identity, collective memory and victimhood in prolonging intergroup/interstate conflicts. Ria has been trained in practical skills and strategies for conflict management and resolution through SIT's CONTACT programme (USA). She is currently the principal investigator of Toda Peace Institute's research project exploring the nexus between climate change, migration, loss of land and identity among youth in the Pacific diaspora communities in Aotearoa New Zealand, Australia and Fiji (Rabi and Kioa islands).

Acronyms

ABG	Autonomous Bougainville Government
ACF	Adaptation Coalition Framework
AIDP	Atoll Integrated Development Policy
ARoB	Autonomous Region of Bougainville
BPA	Bougainville Peace Agreement
CII	Carterets Islands Investments
CoE	Council of Elders
COP	Conference of the Parties
CIRP	Carterets Integrated Relocation Program
EDP	Environmentally Displaced Person
GHG	Greenhouse Gas
GDP	Gross Domestic Product
INGO	International Non-governmental Organisation
IPCC	Intergovernmental Panel on Climate Change
LMMA	Locally Managed Marine Areas
MPA	Marine Protected Area
NGO	Non-governmental Organisation
PICs	Pacific Island Countries
PNG	Papua New Guinea
SIDS	Small Island Developing States
UNDP	United Nations Development Programme
UNFCCC	United Nations Framework Convention on Climate Change
UNSC	United Nations Security Council
UN	United Nations
US	United States of America

1 Introduction. Climate Change and Conflict in the Pacific

Challenges and Responses

Ria Shibata and Seforosa Carroll

It goes without saying that climate change is one of the critical global challenges of our times with dire social, economic, political and environmental consequences. Climate change can have far-reaching social impacts such as potentially inducing conflict and threatening local, regional and global peace and security. Although a direct causal link is yet to be established between climate change and conflict, past studies have analysed and confirmed the role of climate change as a threat multiplier. The economic, social and cultural effects of climate change can generate ripe conditions for conflict over issues such as scarce resources, climate change-induced migration, failed environmental governance or ineffective adaptation and mitigation strategies (Boege 2018). This book argues that climate change policy and practice needs to be conflict-sensitive and it is therefore framed to stress the interconnections between climate change, conflict, security and peace.

The regional focus of this book is the Pacific Islands region. The worst impacts of climate change will occur in low-lying islands and coastal regions, particularly in Pacific Island countries (PICs) where various effects of climate change, including sea level rise, are threatening to displace large numbers of inhabitants. Climate change-induced migration is, therefore, a potential driver of social instability and violence. While there has been a plethora of past research exploring the links between climate change and security, and climate change and violent conflict, both at a global level and in other regions of the world, there has been a significant lack of regional focus on PICs despite their widely acknowledged vulnerability to the conflict-inducing effects of climate change.

The book addresses the various dimensions of the climate change–conflict nexus and sheds light on the overwhelming challenges of climate change in Oceania, for the Pacific peoples in particular. It also highlights the multidimensionality of the problems: political, technical, material, and emotional and psychological. The chapters, written by experts in the field, highlight the centrality and importance of opening up a dialogue between researchers involved in the large-scale global modelling of climate change and its impacts, and experts and civil society actors who are much more knowledgeable about the complexities of local contexts and the conflict-driving potential of climate change

DOI: 10.4324/9781003001744-1

adaptation and mitigation strategies on the ground. The book aims to connect locally focused Pacific approaches with broader international debates in the climate change–security research community. It is our hope that this book will contribute to the development of more conflict-sensitive climate change policies, strategies, governance and adaptation measures in the Pacific region.

Climate Change and Security Threat Discourse

Over the past few decades, climate change has been increasingly discussed as a peace and security issue. Although empirical evidence supporting the 'nexus' between climate change and violent conflict still remains contested among scholars, there is general consensus that climate change may increases risks and existing vulnerabilities that could lead to conflicts. This point was explicitly made in the UN Secretary-General's 2009 report, which described climate change as a 'threat multiplier' that may exacerbate security threats caused by persistent poverty, weak institutions for management of natural resources and conflict resolution, fault lines and a history of mistrust between communities and nations (Report of the UN Secretary-General A/64/350 2009).

Security is considered to be one of the fundamental human needs, an essential requirement for individuals to feel safe and protected in order to thrive mentally and physically (Maslow 1943; Burton 1990). The concept of security may not be primarily about the protection of individual and collective assets from harm, loss and damage. It can also be understood as a shared freedom from fear and want, and the freedom to live in dignity (Ammerdown Group 2016). Climate change can have effects on various forms of security including ontological security (See Campbell Chapter 5; Farbotko 2019, Boege 2022). Ontological security is related to the concept of existential security which refers to the feeling of being secure in one's cultural, religious, political and social identity. Kinnvall and Mitzen (2018) note that ontological security is relational in which everyday interactions and norms are reproduced to create a sense of order, a sense of permanence, continuity and predictability. It is closely linked with the individual's sense of belonging, purpose and meaning and has been regarded as integral to people's mental health and well-being. In this volume, contributors in Parts II and III stress how climate change and relocation of communities threaten the ontological security of the people in the Pacific Islands region.

Climate change is first seen to threaten security of resources. As global temperatures continue to rise and weather patterns become more erratic, water and land resources become scarcer, leading to increased competition and tension between communities and nations. Furthermore, climate change is likely to lead to displacement and migration, as rising sea levels and extreme weather events force people to relocate. This displacement can create new tensions and conflicts as people are forced to compete for resources in newly settled areas. Large-scale mass migration can put additional strain on already fragile social and economic systems, leading to new tensions and conflicts. One example cited by

Buhaug (Chapter 3 in this volume) is the Syrian civil war which has been attributed in part to a severe drought that displaced hundreds of thousands of people and created social and economic tensions that contributed to the eruption of a conflict. Climate change is also likely to exacerbate existing political and social tensions, particularly in regions that are already prone to conflict.

Matt McDonald (2013, 2018, Chapter 2 in this volume) systematically maps out the dominant discourses of climate security, focusing on national security, international security, human security and ecological security. In exploring the contours of climate security discourses, McDonald stresses the importance of understanding the multiple ways in which the relationship between climate change and security is conceived and the need to ask such fundamental questions as: whose security is at stake; who are the key agents of security responsible for responding to the threat; what is the nature of the threat posed by climate change; and what are the responses for dealing with that threat? Analysing the range of diverse voices represented in the climate security discourse is vital in assessing whether these narratives are leading to effective and adequate actions in order to respond to the risks of climate change (McDonald Chapter 2; Barnett and Adger 2007).

The most prominent discourse on climate security has been that which focuses on climate change manifestations and the threat they pose to the security of the nation-state. In this approach, security is defined as preservation of state sovereignty and territorial integrity from external threats (Busby 2018). McDonald points out the limitations of this state-centred securitisation narrative arguing that, "...the causes of climate change are neither acknowledged nor addressed through mitigation efforts; states and their militaries retain their position as central security providers; and victims of climate change are presented as threats to national security" (McDonald Chapter 2). As reflected in US defence and intelligence agency reports (Nuccitelli 2019), flagging climate change as a threat to national security will likely lead to militarisation and reinforcement of national borders to hold back displaced climate migrants.

Another prominent discourse sees climate change as a threat to international security. It emphasises that climate-induced risks to societies and human lives are transnational (Conca 2015; Conca, Thwaites and Lee 2017). Droughts, water shortages, floods, extreme weather events and sea level rise will affect multiple countries in a region at the same time, and the adverse effects are prone to spill over state borders. Hence, this discourse encourages cooperation among states and communities to mitigate climate security risks (Smith and Vivekananda 2007; Busby 2021). McDonald, Boege, Rakova (See Chapters 2 and 9 in this volume) stress the limitations of the international cooperation discourse in failing to address the everyday risks to the health and well-being of vulnerable populations severely affected by climate change consequences.

The human security discourse which highlights the potential threats to individual well-being is promoted by civil society actors and academics (Barnett and Adger 2007). This approach is articulated mainly by those who work to mitigate the climate security risks for vulnerable peoples on the ground. In their chapters, both McDonald and Boege further point out the limitations of a human security-centred approach which fails to address the threats to ecological security or ontological security. The latter in particular goes beyond human beings to all living entities of the cosmic sphere, material and immaterial, including ancestral spirits and future generations. This perspective is particularly relevant when discussing climate change-related threats to which the Pacific Island countries are exposed.

Climate Change and Conflict Nexus

As the adverse risks of climate change to security receive increased attention among global policy makers (Peters 2018), potential causal links between climate change and the risk of violent conflict have been studied extensively over the last decades (Gleditsch 1998; Homer-Dixon 1999; Mach et al. 2019). Previous analyses that have been based on diverse research designs, datasets and methods have resulted in diverse findings. Numerous studies have been conducted to assess whether intensifying climate change will increase the future risk of violent conflict in countries. Compelling results show that the pathways, through which climate change may cause or contribute to conflicts, are complex and highly contextual (Brzoska and Fröhlich 2015; Mach et al. 2019). While there is no strong empirical evidence to demonstrate that climate change causes violent conflict among and within states, there is a growing consensus among experts that it can be a 'threat multiplier' which may act to exacerbate the existing social, political and economic instabilities that communities are already facing (Buhaug 2010). It is clear that tangible effects of climate change such as deviation in temperature, frequent extreme weather events, sea-level rise and increasing resource scarcity can lead to diminished livelihood, economic capacity, and food security and have far-reaching human security implications. Detrimental effects of climate change on livelihoods [such as agriculture, livestock and fisheries] are seen to be one of the key pathways from climate change to violent conflict (Mobjörk and van Baalen 2016). Threats to livelihoods can marginalise affected populations and increase their grievances, leading to greater risk of violence being used to protect or secure the limited resources. Climate-related disasters such as droughts and floods causing soil degradation can also increase people's competition over income-generating resources and lead to potential conflicts. Axbard's study (2016) showed livelihood insecurity from reduction in fishing income to be linked to an increase in piracy activities.

Climate-induced migration is also highlighted as a pathway from climate change to insecurity and conflict (Mobjörk, Krampe and Tarif 2020). Rapid onset of disasters can lead to local displacement but climate change can also result in rural to urban migration or migratory movement to places with better livelihood opportunities (Barnett and Adger 2007). These underlying risks are particularly prevalent in fragile and conflict-affected communities. Fragility increases the vulnerability of communities to climate variability. Vivekananda et al. (2014) warn about the negative cycle between vulnerability and conflict, whereby vulnerability can lead to human insecurity which potentially results in violent conflict which then can further weaken the community's fragility and weak governance. On its own, climate change as a risk factor may be unlikely to cause a violent conflict. However, once climate change interacts with other existing risks such as socio-economic challenges, weak political systems and governance, it can easily become a threat multiplier. The strength of the political systems, governing institutions and actions of the political elites will determine whether nations or communities are able to cope with the additional stress of climate change on their economies and social systems. The threat of conflict therefore increases in fragile states where the pre-existing governance problems are associated with poorly managed resources and low institutional capacity to respond to crises. Climate change can increase grievances by overburdening the coping capacities and lead people to forced migration. The impacts of climate change could diminish the capacity of a government to provide essential services, including public safety and health and food resources, increasing the likelihood of grievances that could lead to violent conflict. The inability of governments to respond adequately to climate-related challenges may lead to reductions in perceived government legitimacy which will lead to the decline of peace and security of the community.

While cumulative environmental variability may not suddenly trigger a violent conflict, a steady build-up of climate-induced challenges coupled with political fragility and ongoing social, political and economic challenges can lead to instability and mobilisation. Hence, if the relationship between climate change and violent conflict is not adequately addressed, there could be a vicious cycle of failure to adapt to climate change, increasing the risk of violent conflict which, in turn, reduces the community's ability to adapt to climate change. This point has been also highlighted in the UN Secretary-General's 2009 report which stresses not only that climate change may serve as 'threat-multiplier' but also that successful adaptation to climate change can serve as a 'threat-minimiser' (UN Report of the Secretary-General 2019).

More and more researchers are shifting the focus on exploring different pathways—both direct and indirect—between climate change impacts and implications for security threats (Mobjörk et al. 2020). Buhaug (Chapter 3) also warns that limited quantitative evidence proving the causal link between climate change and armed conflict should not become a justification for inaction among regional and global policy makers; rather, they should seek measures to support adaptation and minimise security threats imposed by climate change.

Conflict-Sensitive Climate Change Adaptation

Adaptation refers to the adjustment in natural or human systems in response to actual or expected climatic stimuli, or their effect, which moderates harm or exploits beneficial opportunities (IPCC 2007). Adaptation entails a change in processes, practices and structures to prevent or address damages from climate change. The ability to adapt depends greatly on the adaptive capacity or adaptability of an affected system, region, or community to cope with the impacts and risks of climate variability. Adaptation has been described as 'managing the unavoidable,' in contrast to mitigation which is about 'avoiding the unmanageable' (Bierbaum et al. 2007).

Climate change is likely to contribute to conflicts under complex conditions. Purely technical approaches to adaptation will fail to address the plethora of socio-economic, political and other variables that researchers have agreed could contribute to conflict. The authors in this book stress the importance of a holistic, multi-disciplinary approach that includes peacebuilding and socio-economic development perspectives and concerns, and not resorting to technical adaptation responses only, such as building of dams or sea walls. It is therefore integral that adaptation efforts must be conflict-sensitive and be designed to build comprehensive resilience against the impacts of climate change.

Conflict-sensitive adaptation should allow planners and decision-makers to address current vulnerabilities and development priorities, while aiming to ensure long-term sustainability and peace. The various contributions of the authors in this volume have revealed that this type of holistic approach is especially needed when addressing the security threats in the Pacific region.

Climate Change and Security Threats in the Pacific

The Pacific Islands region is comprised of 22 independent political entities and territories that are socially, economically, politically, culturally and linguistically diverse. Although hardly responsible for the world's anthropogenic climatic changes, Pacific Island countries are one of the most vulnerable parts of the world struggling with the disruptive impacts of climate change. This is because many of their islands are low-lying and susceptible to sea-level rise. In addition, extreme weather events such as tropical cyclones have long-lasting effects on the critical infrastructure in Pacific countries' capacities that are not equipped to deal with them. The 2021 World Risk Index's assessment of disaster risk for 181 countries around the world reveals that of all the continents, Oceania has the highest risk due to "high exposure to extreme natural events" (Aleksandrova et al. 2021). The key climatic changes in the Pacific, particularly the small islands, include variability in air and ocean temperatures, ocean chemistry, and rainfall patterns, as well as extreme weather events including tropical cyclones, drought and storms (IPCC 2014). These climate variabilities pose cascading risks to ecosystems and people's livelihoods through greater levels of food insecurity, negative impacts on economic income, wider gaps in gender and youth equality and increasing pressure towards migration or displacement (Crumpler and Bernoux 2020). Climate change disproportionately affects the most vulnerable populations in the Pacific, not only because of exposure to hazards but because of their limited capacity to cope with these climatic risks.

"Climate change remains the single greatest threat to the livelihoods, security, and wellbeing of the peoples of the Pacific" (2018 Boe Declaration on Regional Security).

Climate change is one of the most pressing issues especially affecting the small island states of the Pacific in the face of rapidly transforming environmental conditions. One key area of concern is the impact of climate change on food, water and habitat security of the PICs. Rising sea levels, ocean acidification and coastal erosion, increasing temperatures, changing weather patterns, increased frequency of extreme weather events such as tropical cyclones, floods and droughts, etc. have profound implications for social, economic, political and human security. Among Small Island Developing States (SIDS) in the Pacific, atoll countries such as the Marshall Islands, Tuvalu and Kiribati are viewed as particularly vulnerable (Mimura et al., 2007; Barnett and Campbell 2010; Nurse et al. 2014) The impact of sea level rise threatens to inundate low-lying atoll islands and coastal communities, leading to displacement, forced migration, and conflict over scarce land and resources. A number of studies have highlighted the vulnerability of Pacific Island communities to the impacts of sea level rise, with some projecting that large numbers of people will be forced to relocate in the coming decades. In the Pacific, at least 50,000 islanders are at risk of being displaced by the impact of the disasters and climate change each year (International Organization for Migration 2022).

Rising sea levels not only lead to loss of habitable land and property but can also have long-term impact on the viability of industries such as tourism and agriculture, causing significant threat to economic security. In such ways, changes in weather patterns and resource scarcity can impact water, food and livelihood security, driving social and economic instability and popular grievances. A number of studies have highlighted the potential for climate change to increase risk of conflict in the Pacific, particularly where resources are scarce and competition is high. "The vulnerability (potential for loss) of people to climate change depends on the extent to which people are dependent on natural resources and ecosystem services, the extent to which the resources they rely on are sensitive to climate change, and their capacity to adapt to changes in these resources and services" (Barnett and Adger 2007, 641). The more people are dependent on natural resources for their livelihood security, the greater the risk of climatic changes driving conflict-promoting conditions (Buhaug Chapter 3).

Climate Change, Land and Conflict

The projected changes in climatic conditions pose a severe threat to the Pacific Islanders in the low-lying islands and coastal zones (Nunn 2013). While Pacific peoples have been promoting adaptation efforts, more measures are urgently required as climatic effects intensify. The rise in sea level, coastal inundation and the resulting sea water intrusion will make some of the islands uninhabitable. Migration may be one inevitable response for people whose livelihoods are severely undermined by climate change. Relocation or displacement of communities due to disastrous impacts of climate change nonetheless could lead to competition over scarce resources, which, in turn, could increase the risk of conflict in host communities. Similarly, changes in weather patterns and ocean currents could impact traditional fishing grounds, leading to tensions and competition over limited resources between the migrant community and the host community. When discussing the option of relocation, in Chapter 5, Campbell highlights the importance of understanding the meaning of customary lands to the people of the Pacific.

In addition to the environmental variabilities threatening the Pacific peoples' livelihood and material security, if the residents are forced to migrate, climate change will inevitably have a damaging impact on the people's identity and cultural security. One of the primary challenges of climate change-related migration and relocation is the loss of land and cultural identity. Throughout the Pacific region, people have a special attachment to their traditional lands and resources. There are two dimensions to place attachment for the Pacific Islanders: place identity and place dependence. Both have a large bearing on the extent of connection an individual has with the land (Low 1992). For the Pasifika people, land constitutes the core of their identity. Land is more than just a means of earning a living. Land is deeply connected to the spiritual, emotional, physical and social well-being of the Pasifika people and their communities. Place is not seen merely as a physical location. Land is understood as "a living relational entity with strong spiritual elements which underpin an individual's and group's identity" (Campbell See Chapter 5). Hence, land has both a spatial and temporal meaning that reinforces a sense of belonging and links present and future generations to the ancestors of the past. In this respect, Campbell (2023) and Ravuvu (1983) describe how the Fijian term 'vanua' encapsulates both the past and the potentiality of one's future. Ravuvu stresses that land is life, and "the idea of parting with one's vanua or land is tantamount to parting with one's life" (Ravuvu 1983, 70). The critical importance of land to the Pacific peoples' cultural identity is also expressed in the Samoan word 'fanua' which means both land and placenta (Vaai 2019). And yet the inseparable connection that the Pacific people have to their land will become inevitably threatened by climate change-induced migration and relocation.

Kempf and Hermann cite the case of the Banabans resettlement in Fiji's Rabi Island in 1945 as an example where lack of place and belonging led the residents to experience a sense of anxiety and fear about potential for conflict with other cultural groups in Fiji. For several decades after the relocation, Banabans expressed their emotion in the word 'raraoma' which means insecurity, uncertainty and worry (Kempf and Hermann 2014, 194).

> The first point to make is that *raraoma* has long dominated how Banabans relate to their island of origin, since they were extremely worried as to whether the rights based on belonging to the latter would

remain in force after Banaba was incorporated into the state of Kiribati. The second is that the rise of ethno-nationalism in multicultural Fiji has caused Banabans to fear that full recognition of belonging to the present postcolonial state of Fiji might be withheld, along with the rights and entitlements deriving from such membership.

(Hermann 2004, 211–212)

This example of Banaban resettlement demonstrates that mass relocation of communities, whether as an adaptive measure or a last resort, becomes associated with different materials as well as non-economic loss and damage that increase the vulnerabilities of the communities and become a potential driver of conflict (Kempf and Hermann 2014).

For Pacific Island communities, separation from their ancestral lands and relocation to new areas can entail various risks which include non-material losses such as loss of traditional knowledge and customary practices that are deeply connected to the natural environment. "Losses to ecosystem services, environmental biodiversity due to climate change is closely linked to loss of land and have cascading effects on Pacific people's livelihoods, indigenous knowledge, ways of life, well-being, culture and heritage" (Westoby et al. 2022). Furthermore, McNamara et al.'s (2021) systematic study of non-economic loss and damage in the Pacific revealed that non-economic loss and damage induced by climate change in the Pacific Islands region is closely associated with "fears of cultural loss, deterioration of vital ecosystem services, and dislocation from ancestral lands, among others." The importance of maintaining Indigenous knowledge as 'a way of life and being for Pacific Islanders' is also stressed in Kate Higgins' case study on climate change as a conflict multiplier in Solomon Islands. She argues that "[c]onnection with one's environment is a key aspect of the relational worlds in which peace and security are governed in Solomon Islands. Peace and justice are embedded in beliefs, practices, processes and institutions woven into landscapes and seascapes" (See Higgins Chapter 7). Findings in McNamara et al.'s study reinforce Higgins' point that the non-material damages associated with loss of land are deeply interconnected for the Pacific Island people:

> Land, Family (Home), Spirituality, and Culture (Identity) are what makes a person whole. Remove any one of these elements and the equilibrium will be tipped or swayed to one side more than another causing an imbalance in how things are played out in society (participant #15)
>
> (McNamara et al. 2021, 8)

These studies show that loss of access to ancestral lands due to climate-induced relocation needs to be taken seriously when planning adaptation (e.g., Adger et al. 2011). Climate change migration and displacement can lead to dangerous fragmentation of communities, loss of social structures and relationships. Traditional knowledge and practices are passed down through Pacific social networks. Loss of these networks can also result in loss of cultural identity and Indigenous knowledge. In such ways, relocation can lead to multiple cascading effects threatening not only livelihoods, but also the social and ontological security, emotional and spiritual well-being of Pacific Island people. Hence, the fear of losing connection to one's land is a significant factor that contributes to Pacific peoples' resistance to migration as an adaptation option in the face of climate change. Migration is often seen as a last resort when in situ adaptation measures fail.

Adaptation, Migration and Conflict Risks in the Pacific

While Pacific Islanders have a long and established history of migration, Boege stresses that climate change poses a different kind of existential threat. Climate change consequences may eventually lead to the uninhabitability of the islands, giving involuntary migrants a no return option (Boege Chapter 4). He argues that particularly for low-lying atolls, options for in situ adaptations like building sea walls or planting mangroves are costly, and only work as temporary measures. One risk factor that may lead to conflict when Pacific communities have to be relocated or displaced due to climate change is related to land and resource management. Campbell (Chapter 5) rightly stresses that the essential link between Pacific Island people and their land could arise as the major challenge not only for those forced to leave their land but also for communities within the region that may be required to give up their land for relocatees. Boege and Rakova (Boege and Rakova Chapter 6) highlight the issue of land security and resource governance as a challenge in the case study of the relocation programme of Carteret Islanders from their atoll to mainland Bougainville in Papua New Guinea. Land and access to its resources form an integral part of the local community's identity, with a strong link to its history, sense of belonging and indigenous practice. Because the systems by which customary land is managed and distributed is complex in the Pacific, this makes it increasingly difficult for incoming settlers to negotiate acquisition of customary land in their new environments. Climate-related conflict risk in the Pacific region is not about armed conflicts or violent wars. Boege cautions that we need to be aware of everyday conflicts in local contexts that can escalate to use of violence between communities over scarcity of land and resources, discrimination against incoming climate migrants, or domestic violence against women and children (Boege Chapter 4). The case study by Boege and Rakova (Boege and Rakova Chapter 6) reveals that disputes over land with the recipient communities have forced some Carteret Islanders to return to their home islands. They point out that this kind of localised conflict often remains under the radar of global empirical research and is dismissed.

All the contributors to the Pacific section (see Boege, Campbell, Higgins and Davies in this volume) call for the need to enable Pacific Islanders to exercise agency and adapt to the impacts of climate change on their own terms (see also Bryant-Tokalau 2018). They emphasise the need to understand

and incorporate Pacific peoples' cosmology and the concept of relationality (Vaai 2019) when planning climate adaptation. The need to incorporate Indigenous knowledge is also highlighted in Davies' case study of climate change and conflict prevention in Vanuatu. She encourages the use of *kastom* which is a hybrid of customary laws, traditional ecological knowledge and way of life as a framework for negotiating localised climate adaptation solutions that are both culturally appropriate and empowering for the affected communities (Davies Chapter 8). At the same time, in her case study of Solomon Islands, Higgins (Higgins Chapter 7) also advocates that external interveners from regional and international institutions should trust in the existing local adaptive capacities and the communities' ability to manage their own conflicts. She points out that top-down, state-led international interventions often fail to promote meaningful adaptive programmes because they fail to work with existing customary forms of leadership like chiefs and churches and fail to include women and youth in the conversations. The Pacific case studies in this book underscore the importance of an inclusive process of climate change mitigation and adaptation that involves all levels of stakeholders and recognises the centrality of local ontologies and traditional ways of life.

The contributions in this volume introduce the important experiences and local perspectives of the Pacific people. We hope that the insights shared by the authors will lead to conflict-sensitive and meaningful climate change adaptation and interventions in the future.

Outline

This book is a dialogue that brings together different approaches and ontologies in an integrated and holistic way. It attempts to link the critical threads in the global discourse on the climate change–security–conflict nexus with local Pacific perspectives on culturally appropriate ways of adapting to climate change challenges with dignity. Part I of the book introduces the current global discourse on climate change and security and the contested literature on the climate change–conflict nexus.

In Chapter 2, Matt McDonald uses a critical studies framework to map out the disparate approaches and perspectives in the current discourse on the climate change–security nexus. He argues that there are multiple ways of conceiving the relationship between security and global climate change with different perceptions of who is in need of being protected, from what threat, by which actors, and through what means. Understanding how the association between climate change and security is viewed and by whom it is being advocated is particularly important because it relates to the kind of actions and practices that will follow from different approaches. For example, McDonald suggests that the national and international security discourses which have become prominent in popular consciousness and institutionalised in state and intergovernmental institutions focus on the protection of international order or sovereignty and territorial integrity of nation-states, and fail to effectively address the threats experienced by those directly affected and displaced by climate change. McDonald warns that discourses that are oriented towards the preservation of the status quo may be insufficient when considering the scale of existential threat that climate change poses to our world.

In Chapter 3, Halvard Buhaug presents the contested academic debate on whether climate change is a source of social instability and conflict. He asks the central question, "It is clear that climatic events can have immediate impacts on human security (health, livelihood, food security) but does climate change also constitute a direct threat to peace and societal stability?" He argues that, while there is a lack of clear scientific evidence proving that climate change can cause violent conflict among and within the states, there is an increasing consensus among scholars that climate change can aggravate existing security challenges. In this chapter, he discusses the three components that are relevant to the global climate security debate: (i) the concentration of armed conflict in environmentally fragile regions; (ii) the scientific evidence for the causal relationship between adverse climatic changes and armed conflict; and (iii) the role of climate-related security threats in a comparative perspective. He makes an important argument that a lack of robust evidence on the direct causal link between climatic variability and violent conflict should not diminish the criticality of the global effort at mitigation and adaptation. Climate change will have serious effects on human security in already fragile communities with weak governance structure. He warns that this type of discourse can very well justify inaction by actors who may use it to evade responsibility.

Part II highlights the implications of climate change on the security of the Pacific Island countries and the potential risk of climate-related conflicts. These important case studies explore the nexus between climate change–security threats and conflict, highlighting the critical issues that will affect the Pacific Island people in a variety of ways.

In Chapter 4, Volker Boege lays the overall landscape by calling attention to the various vulnerabilities that Pacific Island countries are facing, such as extreme exposure of coastal regions, frequency of extreme weather events, reduced natural resources and their constrained options for adaptation. Boege stresses that the threats to Pacific Island peoples' security are not limited to economic and physical threats; social and cultural issues also loom large. He calls attention to the potential sources of conflict such as competition over available land, scarce natural resources, conflicts arising from climate-driven migration and relocation, and grievances caused by weak governance and poorly designed adaptation measures. Boege's chapter explores the nexus of climate change and social effects, and their implications for conflict and security in the Pacific. This chapter makes an important contribution to the climate change–conflict discourse in the Pacific by presenting options for conflict-sensitive governance, highlighting the need to involve non-state customary actors and institutions as well as incorporating indigenous cultural approaches to climate change adaptation. Boege calls for an integrated and holistic, multi-scalar approach that involves different actors and organisations to develop an effective governance framework to respond to climate change issues.

In Chapter 5, John R. Campbell explores the importance of land to the Pacific Island people, and reinforces Boege's points about the strong linkages between climate change, land, migration and potential conflict. Campbell examines the meaning of land in the Pacific by illustrating Pacific peoples' deep connections with their customary lands. He argues that "the people and their land are mutually constituted and…one cannot be considered complete without the other." Climate change poses two broad problems in relation to this union. Campbell then describes the likely and already occurring impacts of climate change on the land, focusing on how climate change may damage the land and its ability to support its people. Further, community relocation and displacement of customary lands as an extreme form of adaptation to climate change may have serious impacts on the Pacific people's emotional and spiritual well-being. Where there has been significant in-migration to areas in the region, tensions and conflict have often arisen, frequently with land as a significant underlying issue. Campbell concludes that finding durable solutions for climate change migrants is likely to be a critical issue in the future.

Finally, through three compelling case studies, Part III brings together the findings of climate change experts and the voices of local peacebuilders dealing with conflict-sensitive climate change strategies on the ground. These case studies highlight the conflict-prone effects of climate change, and the criticality of considering conflict-sensitive climate change adaptation and mitigation measures, which draw on the local people's adaptive capacities.

In the first case study in Chapter 6, Volker Boege and Ursula Rakova presents the realities of one of the prominent cases of resettlement in the Pacific region—the relocation of people from the Carterets atoll—part of the Autonomous Region of Bougainville, Papua New Guinea—to the main island of Bougainville. This chapter introduces the specific case of the Carterets Integrated Relocation Program and the various challenges faced by Tulele Peisa, the non-governmental organisation (NGO) responsible for implementing the relocation programme. Boege introduces a wide range of issues that emerged when Carteret Islanders resettled into a fragile region with a history of violent conflicts. He, too, calls special attention to the issue of the land, focusing on the problem of land scarcity and the difficulty of negotiating a just acquisition of customary land between the relocating community and the landholders of the host community. Boege therefore calls for long-term planning and early involvement of the affected community members when planning climate change relocation which recognises the agency, rights and dignity of the relocated families.

The second case study is presented by Kate Higgins in Chapter 7. She examines the climate change and conflict nexus through three elements of the Solomon Islands peace and conflict context—in relation to environmental degradation exacerbating conflict, patterns of internal migration and the risks associated with increased urbanisation, and third, external climate-intervention (including state interventions)—that can potentially constitute a form of climate change-related conflict. Taking seriously the relational and interdependent nature of Pacific communities, Higgins reinforces the importance of indigenous leadership in local adaptation methods, conflict prevention and peacebuilding. She stresses the necessity of indigenous leadership, and the ongoing negotiation of social, political, economic and spiritual forms of power, which she argues is often missing in externally led state and international responses to climate change. She concludes that addressing the climate change and conflict nexus requires a shift in focus that pays more attention to the *relationship* between communities and the state and maps out the consequences of significant structural change for the ways in which interventions can be carried out.

In Chapter 8, Kirsten Davies presents a case that demonstrates the social and conflict-prone implications of climate change in Vanuatu, and strongly argues for the importance of community-led, localised approaches to climate change adaptation, particularly for small, developing states in the Pacific region. Davies argues food insecurity from changing climatic pattern and inundation of land due to sea level rise serve to disproportionately increase the vulnerabilities of Vanuatu on all fronts: geographic, governance, socio-economic and infrastructure. She further warns that forced relocation and displacement of Ni-Vanuatu people driven by climate-induced threats can potentially lead to conflict. When considering options to mitigate the likely consequence of such conflict, Davies strongly advocates for the need to incorporate *kastom* which encapsulates traditional customary law and knowledge as a culturally sensitive platform when developing climate adaptation measures in Vanuatu. She feels that incorporation of traditional, Indigenous knowledge (*kastom*) and empowerment of local leadership (Chiefs) are vital in ensuring the planning and implementation of conflict-sensitive climate responses that engage the local communities.

Lastly, in his concluding chapter, Volker Boege summarises the key threads that emerge in the volume, stressing that Pacific-focused research can make a significant contribution to the wider global debate on linkages between climate change, security and conflict. He argues that Pacific perspectives on climate change and its consequences remain marginalised in the global climate security discourse. He urges policy makers to take into consideration the everyday security impacts at the local level, in particular, the importance of giving recognition to indigenous ways and the involvement of a broad spectrum of actors and local systems of governance to ensure culturally sensitive, and conflict-preventative climate adaptations. This book brings together insights from the Global North alongside local perspectives and the lived experiences of the Pacific Island countries and peoples. We hope it bridges current gaps and contributes to a more comprehensive discourse on climate change, security and conflict so that policy makers in Oceania and elsewhere can develop more grounded and effective responses and actions.

References

Adger, W. Neil, Jon Barnett, F. Stuart Chapin III, and Heidi Ellemor. 2011. "This Must Be the Place: Underrepresentation of Identity and Meaning in Climate Change Decision-Making." *Global Environmental Politics* 11 (2): 1–25. https://doi.org/10.1162/GLEP_a_00051

Aleksandrova, Mariya et al. 2021. *WorldRiskReport 2021*. Bündis Entwicklung Hilft and Ruhr University Bochum – Institute for International Law of Peace and Armed Conflict (IFHV). https://www.ifhv.de/publications/world-risk-report

Ammerdown Group. 2016. *Rethinking Security: A Discussion Paper*. May 2016.

Axbard, Sebastian. 2016. "Income Opportunities and Sea Piracy in Indonesia: Evidence from Satellite Data." *American Economic Journal: Applied Economics* 8 (2): 154–194.

Barnett, Jon, and W. Neil Adger. 2007. "Climate Change, Human Security and Violent Conflict." *Political Geography* 26 (6): 639–655.

Barnett, Jon, and John R. Campbell. 2010. *Climate Change and Small Island States: Power, Knowledge and the South Pacific*. London and Washington, DC: Earthscan.

Bierbaum, Rosina et al. 2007. *Confronting Climate Change: Avoiding the Unmanageable and Managing the Unavoidable*. Scientific Expert Group Report on Climate Change and Sustainable Development. United Nations Foundation.

Boe Declaration. 2018. *Forty-Ninth Pacific Islands Forum, Boe Declaration on Regional Security*. 6 September 2018. https://www.forumsec.org/boe-declaration-on-regional-security/

Boege, Volker. 2018. "Climate Change and Conflict in Oceania: Challenges, Responses, and Suggestions for a Policy-Relevant Research Agenda." Toda Peace Institute Policy Brief, no. 17. Tokyo: Toda Peace Institute.

Boege, Volker. 2022. "Ontological Security, the Spatial Turn and Pacific Relationality Part I." Toda Peace Institute Policy Brief, no. 123. Tokyo: Toda Peace Institute.

Bryant-Tokalau, Jenny. 2018. *Indigenous Pacific Approaches to Climate Change: Pacific Island Countries*. Cham, Switzerland: Palgrave Macmillan.

Brzoska, Michael, and Christiane Fröhlich. 2015. "Climate Change, Migration and Violent Conflict: Vulnerabilities, Pathways and Adaptation Strategies." *Migration and Development* 5 (2): 190–210.

Buhaug, Halvard. 2010. "Climate Not to Blame for African Civil Wars." *Proceedings of the National Academy of Sciences* 107 (38): 16477–16482.

Burton, John W. 1990. *Conflict: Human Needs Theory*. New York: St. Martin's Press.

Busby, Joshua W. 2018. "Warming World." *Foreign Affairs* 97 (4): 49–55.

Busby, Joshua W. 2021. "Beyond Internal Conflict: The Emergent Practice of Climate Security." *Journal of Peace Research* 58 (1): 186–194.https://doi.org/10.1177/0022343320971019

Conca, Ken. 2015. *Un Unfinished Foundation. The United Nations and Global Environmental Governance*. Oxford: Oxford University Press. https://doi.org/10.1093/acprof:oso/9780190232856.001.0001

Conca, Ken, Joe Thwaites, and Goueun Lee. 2017. "Climate Change and the UN Security Council: Bully Pulpit or Bull in a China Shop?" *Global Environmental Politics* 17 (2): 1–20.

Crumpler, Krystal, and Martial Bernoux. (2020). "Climate Change Adaptation in the Agriculture and Land Use Sectors: A Review of Nationally Determined Contributions (NDCs) in Pacific Small Island Developing States (SIDS)." In *Managing*

Climate Change Adaptation in the Pacific Region, edited by Walter Leal Filho, 1–26. Cham, Switzerland: Springer Nature.

Farbotko, Carol. 2019. "Climate Change Displacement: Towards Ontological Security." In *Dealing with Climate Change on Small Islands: Towards Effective and Sustainable Adaptation?*, edited by Carola Kloeck, and Michael Fink, 251–266. Goettingen: Goettingen University Press.

Gleditsch, Nils Petter. 1998. "Armed Conflict and the Environment: A Critique of the Literature." *Journal of Peace Research* 35 (3): 381–400.

Hermann, Elfriede. 2004. "Émotions, Agency and Displaced Self of the Banabans in Fiji." In *Shifting Images of Identity in the Pacific*, edited by Toon van Meijl and Jelle Miedema, 191–217. Leiden: KITLV Press.

Homer-Dixon, Thomas F. 1999. *Environment, Scarcity, and Violence*. Ewing: Princeton University Press.

Intergovernmental Panel on Climate Change (IPCC). 2007. *Climate Change 2007: Synthesis Report. Contribution of Working Groups I, II and III to the Fourth Assessment Report of the Intergovernmental Panel on Climate Change* [Core Writing Team, Rajendra K. Pachauri and Andy Reisinger. (eds.)]. IPCC, Geneva, Switzerland, 104 pp.

International Organization for Migration (IOM). 2022. Pacific Response to Disaster Displacement (PRDD). https://environmentalmigration.iom.int/pacific-response-disaster-displacement-prdd

IPCC 2014. *Climate Change 2014: Synthesis Report. Contribution of Working Groups I, II and III to the Fifth Assessment Report of the Intergovernmental Panel on Climate Change* [Core Writing Team, Rajendra K. Pachauri and L.A. Meyer (eds.)]. IPCC, Geneva, Switzerland, 151 pp.

Kempf, W., and E. Hermann. 2014. "Epilogue. Uncertain Futures of Belonging: Consequences of Climate Change and Sea-Level Rise in Oceania." In *Belonging in Oceania: Movement, Place-Making and Multiple Identifications*, edited by E. Hermann, T. van Meijl, and W. Kempf, 189–213. Oxford: Berghahn.

Kinnvall, Catarina, and Jennifer Mitzen. 2018. "Ontological Security and Conflict: The Dynamics of Crisis and the Constitution of Community." *Journal of International Relations and Development* 21 (4): 825–835.

Low, Setha M. 1992. "Symbolic Ties That Bind." In *Place Attachment*, edited by Irwin Altman and Setha M. Low, 165–185. New York: Plenum Press.

Mach, Katharine J., Caroline M. Kraan, W. Neil Adger, Halvard Buhaug, Marshall Burke, James D. Fearon, Chris B. Field, Cullen S. Hendrix, Jean-Francois Maystadt, John O'Loughlin, Philip Roessler, Jürgen Scheffran, Kenneth A. Schulz, and Nina von Uexkull. 2019. "Climate as a Risk Factor for Armed Conflict." *Nature* 571: 193–197.

Maslow, Abraham H. 1943. "A Theory of Human Motivation." *Psychological Review* 50(4): 370–396.

McDonald, Matt. 2013. "Discourses of Climate Security" *Political Geography* 33: 42–51.

McDonald, Matt. 2018. "Climate Change and Security: Towards an Ecological Security Discourse?" *International Theory* 10 (2): 153–180.

McNamara, Karen, Ross Westoby, Rachel Clissold, and Alvin Chandra. 2021. "Understanding and Responding to Climate-driven Non-economic Loss and Damage in the Pacific Islands." *Climate Risk Management* 33 (2021) 100336: 1–14.

Mimura, Nobuo, Leonard Nurse, Roger F. McLean, John Agard, Lino Briguglio, Penehuro Lefale, Rolph Payet, and Graham Sem. 2007. "Small Islands." In *Climate*

Change 2007: Impacts, Adaptation and Vulnerability, edited by Martin L. Parry, Osvaldo F. Canziani, Jean P. Palutikof, Paul J. van der Linden, and Clair E. Hanson, 687–716. Contribution of Working Group II to the Fourth Assessment Report of the Intergovernmental Panel on Climate Change. Cambridge: Cambridge University Press.

Mobjörk, Malin, and Sebastian van Baalen. 2016. "Climate Change and Violent Conflict in East Africa—Implications for Policy." SIPRI Policy Brief, April 2016.

Mobjörk, Malin, Florian Krampe, and Kheira Tarif. 2020. "Pathways of Climate Insecurity: Guidance for Policymakers." SIPRI Policy Brief, November 2020.

Nuccitelli, Dana. 2019. *Climate Change Poses Security Risks, According to Decades of Intelligence Reports: The Long History of Climate Change Security Risks.* Yale Climate Connections. https://www.yaleclimateconnections.org/2019/04/the-long-history-of-climate-change-security-risks/

Nunn, Patrick. 2013. "The End of the Pacific? Effects of Sea Level Rise on Pacific Island Livelihoods." *Singapore Journal of Topical Geography* 34 (2): 143–171. https://doi.org/10.1111/sjtg.12021

Nurse, Leonard A., Roger F. McLean, John Agard, Lino Pascal Briguglio, Virginie Duvat-Magnan, Netatua Pelesikoti, Emma Tompkins, et al. IPCC 2014. "Small Islands". In *Climate Change 2014. Impacts, Adaptation, and Vulnerability. Part B: Regional Aspects.* Contribution of Working Group II Contribution to the Fifth Assessment Report of the Intergovernmental Panel on Climate Change, edited by Vicente R. Barros, Christopher B. Field, David Jon Dokken, Michael D. Mastrandrea, Katharine J. Mach, T. Eren Bilir. Cambridge, UK and New York, NY: Cambridge University Press.

Peters, Katie. 2018. "Disaster, Climate Change, and Securitisation: The United Nations Security Council and the United Kingdom's Security Policy." *Disasters (Special Issue: Disasters in Conflict Areas)* 42 (S2): 196–214.

Ravuvu, Asesela. 1983. *Vaka i taukei. The Fijian Way of Life.* Suva: University of the South Pacific.

Smith, Dan, and Janani Vivekananda. 2007. *A Climate of Conflict: The Links between Climate Change, Peace and War.* London: International Alert.

UN Report of the Secretary-General. 2019. Document A/64/350.

Vaai, Upolu Luma. 2019. "We Are Therefore We Live: Pacific Eco-relational Spirituality and Changing Climate Change Story." Toda Peace Institute Policy Brief, no. 56, October 2019.

Vivekananda, Janani, Janpeter Schilling, and Dan Smith. 2014. "Climate Resilience in Fragile and Conflict-Affected Societies: Concepts and Approaches." *Development in Practice* 24 (4): 487–501. https://doi.org/10.1080/09614524.2014.909384

Westoby, Ross, Rachel Clissold, Karen E. McNamara, Anita Latai-Niusulu, and Alvin Chandra. 2022. "Cascading Loss and Loss Risk Multipliers Amid a Changing Climate in the Pacific Islands." *Ambio* 51: 1239–1246. https://doi.org/10.1007/s13280-021-01640-9

Part I

Climate Change-Security-Conflict Nexus

2 The Climate Change–Security Nexus

A Critical Security Studies Perspective

Matt McDonald

Introduction

Climate change is increasingly recognised as a security issue. It has been discussed at the UN Security Council, it features in national security strategy documents of more than half of the world's states, and a wide range of think tank and academic publications point to the intersection between climate change and security.

This does not mean, however, that there is consensus about the climate–security relationship or the desirability of linking the two. Some theorists working in the broad tradition of critical security studies were important voices in pointing to the security implications of climate change, while others (perhaps paradoxically) urged caution in linking climate and security. In this sense it's fair to say there's no single 'critical security studies' perspective on the climate–security nexus, just as there is no single 'critical security studies' perspective in general.

Critical security studies can be defined (broadly) as scholarship concerned with developing a critique of traditional approaches to security; examining the politics of security and exploring the ethical assumptions and implications of particular security discourses and practices (see Browning and McDonald 2013).

This chapter provides an introduction to what Critical Security Studies has to offer in understanding and guiding practice on the climate change–security nexus. It suggests that analysis consistent with the critical security studies tradition can be (and has been) brought to bear on the climate change–security nexus by examining the scope of security threats; exploring the contested meanings of 'climate security' and engaging key questions and dilemmas associated with linking the two, in theory and practice. It also provides a brief illustration of the utility of a critical security studies perspective when approaching the relationship between climate change and armed conflict, and concludes with policy recommendations.

Broadening Security to Include Climate Change

Critical approaches to security were at the heart of attempts in international relations thought to challenge dominant accounts of security that focused on the territorial preservation of the nation-state from external military threat. This

DOI: 10.4324/9781003001744-3

dominant conception of security was advanced by Realists, with security—defined as state survival—in turn, seen as the central goal and ambition of states. For Realists, the international system was defined by the absence of a higher authority than states. This anarchic environment encouraged states to concern themselves with their own survival, and encouraged them to view the motives and actions of other states with suspicion. With no higher authority to prevent conflict or regulate the international system, conflict was seen as an inevitable feature of world politics, and preparation for conflict through maximising the state's relative power was necessary to ensure the state's survival. In this context, security was viewed in terms of the state and its survival, and security studies was viewed as 'the study of the threat and use of force' (Walt 1991).

Critical security studies developed critiques of this narrow view of security. The association of security and the state assumed that states were the best means of providing for the welfare of their citizens. But for some critics this account was challenged by the scale of poverty and suffering throughout the world (e.g. Galtung 1969), including suffering generated in many instances by governments themselves (see McSweeney 1999). Feminist scholars, meanwhile, noted that the Realist conception of security arose out of a gendered understanding of security, sovereignty and the state (see, e.g. Tickner 1988). Among other accounts of the limitations of traditional approaches to security, these critiques challenged the exclusive focus on state security and encouraged attention to non-traditional security threats and the welfare of people themselves.

For many advocates of redefining security, it made little sense to exclude existential threats to human survival. In this sense, they argued for including environmental change in an expanded definition of security (Mathews 1989; Renner 1996). And while it was, at face value, an analytical claim about the scope of threats to security, there was also a normative and political element to this position: if environmental issues could be elevated to the 'high politics' of security, they would receive the funding, attention and priority they deserved (see McDonald 2012). More recently, a range of analysts and practitioners have advocated including climate change on states' security agendas for similar reasons.

In the process, advocates for approaching climate change as a security issue have to engage with both the politics and ethics of security. For them, greater political attention to climate change would be desirable (ethics), and elevating the issue to the realm of security would help mobilise a response to the issue by injecting the issue with urgency and priority associated with high politics. Few would argue with the sentiment here, though the political implications of approaching an issue as a security issue divides critical security theorists. This applies as much to climate change as any other issue.

For critics of this move within the critical security studies tradition, two (related) objections have been articulated. The first is that at the heart of the meaning of 'security', we find an association with defence, the military and the state. For most, climate change is ill-suited to this agenda, and some suggest that states and the defence establishment more specifically are embracing this agenda precisely to secure resources and underscore their own legitimacy (see Buxton and Hayes 2015; Marzec 2015). While this objection suggests the *meaning* of security is inevitably tied to the state and defence, the second objection suggests security has a particular *logic*—that defining an issue as a security issue serves to take it out of the realm of deliberation and discussion, instead enabling it to be dealt with via secrecy and illiberal means (Wæver 1995). These related concerns encourage some critical security studies scholars—especially those informed by post-structural thought or employing the 'securitisation' framework—to suggest that climate change would benefit from being decoupled from security. In the process they echo some of the concerns originally advanced by Daniel Deudney in 1990, when he warned of the militarisation of the environment.

For others in the critical security studies tradition, however, the linkage between security and climate change is not *inherently* problematic. For Ken Booth (1991; 2007) for example, the issue here is not with 'security', but with a dominant conception of security tied to the nation-state and its preservation from external threat. Once we escape this definition, we can imagine alternative and more progressive practices to follow, especially if security is oriented to the emancipation of vulnerable populations. Indeed, for Booth and others, it is precisely because security is high politics that it is important not to give up on it as a site of progressive practice, but instead try to shift the way political communities conceive and approach security.

The above accounts clearly have very different implications for conceiving an issue like climate change as a security threat. But both approaches recognise the need to engage with the politics and ethics of security, rather than treat it simply as an abstract analytical category. And both suggest that security is *constructed*, with different communities prioritising different threats and responses to them. The following section expands on these issues through analysing different climate security *discourses*, with their different sets of assumptions and implications.

Climate Change and Discourses of Security

What do those working in critical security studies mean when they talk about *discourses* of security? Simply put, a security discourse is a framework of meaning that conditions the way political communities conceive and approach what is to be protected, by whom, from what threats and by what means (see McDonald 2012; 2013). When looking at the way states, practitioners, analysts and academics engage with the climate change–security relationship, we can identify a number of different discourses, each with a different emphasis on whose security is important.

The first, and arguably most dominant, discourse of climate security is that of national security. Here, a range of analysts and policy practitioners point out that climate change may pose a threat to the sovereignty and territorial integrity of nation-states. Some of the possibilities envisaged here include the potential for state sovereignty to be challenged by large-scale migration of people from other countries displaced by manifestations of climate change, including natural disasters or rising sea levels. Others include interstate competition and even conflict over shared or common resources like freshwater or ocean fisheries, as climate change decreases supply of such resources. Joshua Busby (2018), for example, sees the potential for tension and future conflict over access to the waters of the Indus, the Mekong and the Nile, and over a scramble for resources in the Arctic. Traditional concerns ultimately remain dominant—the sovereignty and territorial integrity of the nation-state from (external) threat.

It is telling in this discourse, embraced by states and think tanks in particular (e.g. Busby 2007; 2008), that measures to respond to these challenges are focused on managing manifestations of climate change rather than causes of the problem; are focused on adaptation rather than mitigation; and conceive of the traditional agent of national security—the state and in particular the military—as the central agent for providing security. Climate change becomes simply another potential cause of instability, violence and warfare with which the state is primarily concerned.

In one extreme example, a 2003 Pentagon report on the national security implications of an abrupt climate scenario for the US suggested that states such as the US may consider building more effective boundaries to prevent those displaced by climate change from entering US territory (Schwartz and Randall 2003). Here, the causes of climate change are neither acknowledged nor addressed through mitigation efforts; states and their militaries retain their position as central security providers; and victims of climate change are presented as threats to national security. In this sense, not least as this is such a prominent discourse of climate security, the concerns of those critical security theorists cautioning against linking climate change and security seem well founded.

Another prominent discourse of climate security is that of international security. This focus was apparent in UN Security Council deliberations on the security implications of climate change from 2007 (Conca, Thwaites and Lee 2017), and arguably better captures the inherently transnational nature of the issue. In this discourse, the focus is on the way in which the international system itself may be challenged through instability caused by climate change. While the focus remains on similar issues to the national security discourse—population movements and (associated) possibility for conflict—the means of addressing these issues extend to cooperative mitigation efforts and conflict prevention involving the UN system, for example (see Purvis and Busby 2004; Smith and Vivekananda 2007).

While this discourse encourages international cooperation in the face of climate change, there are still limitations. In presenting the international status quo as that in need of protection, such an approach fails to systematically engage with the conditions in which climate change itself has become possible.

There is a danger here of focusing on large-scale disruption associated with climate change—mass population movements, regional instability and conflict—rather than the everyday implications for health and economic well-being for those vulnerable populations affected by climate change throughout the world. For those working in the tradition of critical security studies, the international security discourse insufficiently reorients our focus to the key drivers of climate change and its main victims.

An obvious response to the above is a focus on human security, which has become a prominent frame for analysing climate security (see Matthew et al. 2010). Human Security is a discourse advanced principally by NGOs and academics and focuses primarily on the need for cooperative and significant mitigation efforts to minimise harm experienced by vulnerable populations.

Some have gone still further in recognising and endorsing a focus on eco-logical security. In this discourse, the referent objects are ecosystems themselves and their resilience in the face of climate change. This focus enables attention not only to vulnerable populations, but also to other living beings and future generations (see McDonald 2021). Here, radical mitigation efforts oriented to the most vulnerable are most imperative.

Two particular points are worth noting with these latter two discourses. First, while theorists working in the critical security studies tradition have noted the importance of identifying climate security discourses and their ef-fects, some have gone further in specifically *advocating* a focus on human security or ecological security as an appropriate ethical focus. In this sense, they implicitly reject the idea that securitisation is necessarily a bad thing, sug-gesting progressive practices may follow if our security lens is oriented to the wellbeing of vulnerable beings.

Second, however, both confront dilemmas. While perhaps more ethically defensible, these discourses have limited political purchase among and within the key institutions of world politics: states. They require radical practices that orient towards vulnerable outsiders, and in this sense do not appear to have an obvious constituency in the halls of power of key national and international institutions. In advocating such a reconceptualisation of security, then, critical security scholars may be confronted with a dilemma: to what extent should we engage key institutions on their own terms to pursue meaningful change (and risk these institutions focusing predominantly on their own survival), and to what extent should we outline radical alternatives even in the face of the appar-ently contradictory interests of powerful actors (and risk political irrelevance)?

Climate Change and Security: Key (Critical) Questions

The above analysis suggests significant points of disagreement between schol-ars working within the broad tradition of critical security studies. But it also suggests key questions in engaging the climate–security relationship.

The first of these, following from the above discussion of climate security discourses, is 'How exactly is security understood, and in particular whose security is presented as threatened by climate change?' A critical approach to the climate–security relationship recognises the constructed nature of security, and the political choices involved in prioritising certain sets of threats and downplaying others, for example. In this context, examining exactly how this relationship is understood and approached is important, especially if we accept the view that these choices have important implications for approaching climate change in practice.

Three further key questions arise from this.

1 First, 'what are the effects of the securitisation of climate change for policy and practice?'.

   As noted, the securitisation framework suggests that constructing climate change as a security issue will usher in 'panic politics' and forms of extraordinary practices inconsistent with political deliberation and debate (Wæver 1995; Buzan, Wæver, and Wilde 1998). While not all critical security studies scholars would embrace this view, all recognise the importance of engaging with the effects of representing and approaching climate change as a security issue. For them this is not simply an abstract analytical exercise—it has practical significance.

2 Second, 'how *should* we define and approach the relationship between climate change and security?'.

   Again, scholars working across the critical security studies tradition advance very different answers to this question. These range from the imperative of escaping a security frame to the need to recognise climate change as the most pressing global and ecological security threat. But all identify the need to engage with this normative/ ethical question about the climate–security relationship, recognising that ethical questions are fundamental to both the study and practice of security.

3 Third, 'why do different political communities understand and approach the climate change–security relationship in particular ways?'.

   Simply put, how do we make sense of the construction of climate security in particular contexts and of the way particular approaches to this relationship have 'won out' over others. A quick analysis of UN Security Council debates over the international security implications of climate change from 2007, for example, reveals significant disagreement over whether it is appropriate to discuss climate change in the Security Council at all, and if so how the relationship between international security and climate change should be understood (see Maertens 2021; Scott 2008; Conca, Thwaites and Lee 2017). The same is arguably true within states, with different actors advancing alternative accounts of whose security is in need of protection and how it should be protected. For critical approaches, security itself is a site of contestation. While an important analytical point, this is also an important question for normative and political reasons. Understanding how certain ideas about the climate change–security relationship (and associated practical responses to climate change) come to prominence in certain settings can provide us with insights into the conditions in which alternative approaches might come to prominence.

Beyond these questions, feminist theorists have drawn our attention to the role of gender in conditioning how we think about the relationship between climate change and security (e.g. Detraz 2009); critical political geographers to the role of representations of space in climate security discourses (e.g. Dalby 2009) and postcolonial scholars to the centrality of Western thought and experience to existing (dominant) accounts of the climate change–security relationship (e.g. Grove 2010). While by no means an exhaustive list, these accounts all draw attention to neglected dimensions of the climate change–security relationship, raising important questions about how we view this approach in theory and in practice. And in combination with the preceding questions, those working in the critical security studies tradition compel us to reflect on the limits and implications of traditional approaches to climate change and security, the politics of linking climate and security, and the ethical assumptions and implications associated with this linkage.

Case Study: Climate Change and Conflict

Much of the above account suggests the need to step back and consider the relationship between climate change and security in fundamental terms. But how would scholars working in this tradition engage a substantive issue like the relationship between climate change and conflict? This is potentially challenging in the sense that this concerns an issue central to the traditional security agenda: armed conflict. This has been a broader issue for critical security theorists, with Ken Booth (1991) among others making the case that we need to bring the resources of this tradition to bear to better understand conflict, and guide more appropriate and effective responses to it. War, he argues, is too important to be left to strategists.

A range of analysts have suggested recently that climate change could serve as a driver of conflict, or has already played a role in triggering armed conflict in different settings (see Smith and Vivekananda 2007 or Busby 2018, for example). Three arguments have been made here—three envisaged pathways between climate change and armed conflict.

1 In the first argument, changes in rainfall pattern, increasing temperatures and desertification associated with climate change may change previously arable land into land that cannot support local communities. In response to this we may see population movements, with groups of people coming into contact and ultimately conflict with other groups. This argument was invoked in the case of conflict in Darfur, with then UN Secretary-General Ban Ki Moon (2007) and the UN Environmental Program (2007) both pointing to the role of climate change in contributing to this conflict. We also see a similar logic in arguments suggesting the possibility of future

conflict as populations are displaced by rising sea levels or natural disasters linked to climate change.

2 The second argument is one that suggests that some of the same manifestations of climate change—changing rainfall, loss of arable land and desertification—may trigger political unrest within states. The argument here is that a population struggling with insufficient access to food, freshwater or means to make a living will be more likely to confront their government for a greater or more equitable share of national resources, especially if there are existing grievances, weak institutions or social/ political tensions. This argument has been applied to civil war in Syria, the wider Arab Spring and violence in Mali (see CNA 2014; Dunlop and Spratt 2017, 13; Selby et al. 2017).

3 The third argument suggests that a climate-induced increase in demand and decrease in supply for a resource (like freshwater, for example) will lead to accelerated resource extraction and/ or manipulation, triggering contestation and even conflict. This pathway has been linked to future transboundary water wars, and conflict over access to resources found in the global 'commons', such as fish stocks in oceans (see Gleick 1993; Miller 2000; Busby 2018).

So, what would scholars working in the critical security tradition have to say about the above conflicts and arguments, and what would they contribute to allow us to better understand and respond to these instances of violence and/ or climate-related stress? I would identify four key contributions that analysts working in this tradition could bring to bear on the climate-conflict relationship.

1 *Avoid simplistic accounts of causation.* Critical approaches would discourage simple associations between (climate induced) stress and violence. In theoretical terms, their rejection of reductionist and parsimonious accounts of security extends to the question of why actors engage in violence. While Realists would suggest that competition over resources will inevitably drive conflict, analysts working in the critical tradition would be more likely to acknowledge the possibility of cooperation and/or the wide range of factors that create conditions for conflict. At best, climate change serves in this context as a 'threat multiplier': creating contexts in which conflict is more likely, but not in and of itself *causing* conflict (CNA 2014). This conclusion is one found in many more nuanced accounts of the complex relationship between climate change and violence (e.g. Wolf 1999; Nordas and Gleditsch 2007; CNA 2014; Boege 2018).

2 *Avoid simplistic accounts of actors' motives.* Following from the above, critical approaches discourage assumptions about why actors will behave in certain ways. While some Realists make sweeping claims about the motives of states—suggesting they will always act to maximise power relative to others, with little to prevent the use of force in the process—critical approaches suggest the need for a richer and more nuanced appreciation of

the drivers of actors' behaviour, including that of states. This can range from domestic constraints to international expectations to alternative calculations of their national interests. This, for critical approaches, allows us to recognise that different communities will respond to climate stress in different ways, and allows us to focus on the circumstances in which stress can be met with cooperation and movements towards resilient communities, for example.

3 *Address mitigation as a response.* Traditional approaches to security tend to focus on managing and responding to conflict, less on the circumstances in which conflict becomes possible. While Joshua Busby (2007; 2008) argued that a national security response to climate change should prioritise adaptive measures, critical approaches are far more likely to suggest the need to focus on mitigation. While this is a focus oriented towards the rights and needs of vulnerable populations who will suffer disproportionately from effects of climate change regardless of the likelihood of violence, one effect of this focus is that it orients towards preventing conflict before it occurs rather than responding to it or preparing militarily for managing manifestations of climate stress.

4 *Recognise climate conflict is not the only climate security threat, nor the most important.* With its focus on the territorial preservation of the nation-state from military threat, traditional approaches to security focus our attention on climate-induced conflict as the key threat posed by climate change. Critical approaches, by contrast, suggest that war is not the only nor most important security threat, especially when our attention turns to the impact of climate change on vulnerable populations, future generations or other living beings. Critical approaches to security remind us that the implications of livelihood loss, disease, insufficient food and water or displacement associated with climate change are more immediate, more direct and more pressing threats to the security of people themselves than the danger of armed conflict (see Barnett and Adger 2007).

Ultimately, a critical approach to the climate–conflict relationship is one that calls for caution and nuance in linking climate change and warfare, and recognises that linking the two can potentially draw our attention away from both other causes of conflict and other (more immediately pressing) manifestations of climate change.

Conclusion

Critical approaches to security have much to contribute to our understanding of the relationship between climate change and security. While there is no single 'critical security approach', the range of scholarship in this broad tradition encourages us to reflect on the politics and the ethics of security, in the process adding much to our understanding of the choices made in conceiving and approaching the relationship between climate and security in particular ways.

It also provides us with resources for reimagining and redefining not only the way we approach this relationship, but how we approach the existential threat of climate change itself.

Policy Recommendations

While much of this chapter applies to academic analysis, insights drawing on critical security studies scholarship can also inform practical responses to climate change and security implications associated with it. Key recommendations would include the need for policy practitioners to:

- Develop assessments of climate security risks that are interdisciplinary in nature.
- Develop integrated and holistic responses to climate security threats that promote:
  - Mitigation and adaptation.
  - Action at the level of international cooperation, national policy and sub-national/community policy and practices.
  - Action from traditional security practitioners (e.g. defence forces) along with the wider range of agents able to promote action on mitigation, adaptation and resilience-building.

References

Barnett, Jon, and Neil Adger. 2007. "Climate Change, Human Security and Violent Conflict." *Political Geography* 26 (6): 639–655.

Boege, Volker. 2018. "Climate Change and Conflict in Oceania: Challenges, Responses and Suggestions for a Policy-Relevant Research Agenda." Toda Peace Institute Policy Brief No.17. Tokyo: Toda Peace Institute.

Booth, Ken. 1991. "Security and Emancipation." *Review of International Studies* 17 (4): 313–326.

Booth, Ken. 2007. *Theory of World Security.* Cambridge: Cambridge University Press.

Browning, Christopher S., and Matt McDonald. 2013. "The Future of Critical Security Studies." *European Journal of International Relations* 19 (2): 235–255.

Busby, Joshua. 2007. *Climate Change and National Security.* Washington, DC: Council of Foreign Relations. Available at: ///http://www.cfr.org/publication/14862.

Busby, Joshua. 2008. "Who Cares about the Weather? Climate Change and US National Security." *Security Studies* 17 (3): 468–504.

Busby, Joshua. 2018. "Warming World." *Foreign Affairs* 97 (4): 49–55.

Buxton, Nick, and Ben Hayes. 2015. "Introduction: Security for Whom in a Time of Climate Crisis." In *The Secure and the Dispossessed,* edited by Nick Buxton and Ben Hayes, 1–22. London: Pluto.

Buzan, Barry, Ole Wæver, and Jaap de Wilde. 1998. *Security: A New Framework for Analysis.* Boulder, CO: Lynne Rienner.

CNA. 2014. *National Security Risks and the Accelerating Risks of Climate Change.* Available at: https://www.cna.org/cna_files/pdf/MAB_5-8-14.pdf

Conca, Ken, Joe Thwaites, and Goueun Lee. 2017. "Climate Change and the UN Security Council: Bully Pulpit or Bull in a China Shop?" *Global Environmental Politics* 17 (2): 1–20.

Dalby, Simon. 2009. *Security and Environmental Change.* Cambridge: Polity.

Detraz, Nicole. 2009. "Environmental Security and Gender: Necessary Shifts in an Evolving Debate." *Security Studies* 18 (2): 345–369.

Deudney, Daniel. 1990. "The Case Against Linking Environmental Degradation and National Security." *Millennium* 19 (3): 461–473.

Dunlop, Ian, and David Spratt. 2017. *Disaster Alley: Climate Change, Conflict and Risk.* Melbourne: Breakthrough. Available at https://www.breakthroughonline.org.au/disasteralley

Galtung, Johan. 1969. "Violence, Peace, and Peace Research." *Journal of Peace Research* 6 (3): 167–191.

Gleick, Peter. 1993. "Water and Conflict." *International Security* 18 (1): 79–112.

Grove, Kevin. 2010. "Insuring our Common Future?" *Geopolitics* 15 (3): 536–563.

Maertens, Lucile.2021. "Climatizing the UN Security Council." *International Politics.*58: 640–660.

Marzec, Robert P. 2015. *Militarizing the Environment: Climate Change and the Security State.* Minneapolis: University of Minnesota Press.

Mathews, Jessica. 1989. "Redefining Security." *Foreign Affairs* 68 (2): 162–177.

Matthew, Richard A., Jon Barnett, Bryan McDonald, and Karen L. O'Brien, eds. 2010. *Global Environmental Change and Human Security.* Cambridge, MA: MIT Press.

McDonald, Matt. 2012. *Security, the Environment and Emancipation.* London: Routledge.

McDonald, Matt. 2013. "Discourses of Climate Security" *Political Geography* 33: 42–51.

McDonald, Matt. 2021. Ecological Security: Climate Change and the Construction of Security. Cambridge: Cambridge University Press.

McSweeney, Bill. 1999. *Security, Identity and Interests.* Cambridge: Cambridge University Press.

Miller, Kathleen. 2000. "Pacific Salmon Fisheries." *Climatic Change* 45 (1): 37–61.

Moon, Ban-Ki. 2007. "A Climate Culprit in Darfur." *Washington Post*, 16 June, 2007.

Nordas, Ragnhild, and Nils Gleditsch. 2007. "Climate Change and Conflict." *Political Geography* 26 (6): 627–638.

Purvis, Nigel, and Joshua Busby. 2004. "The Security Implications of Climate Change for the UN System." *Environmental Change and Security Project Report* 10: 67–73.

Renner, Michael. 1996. *Fighting for Survival: Environmental Decline, Social Conflict and the New Age of Insecurity.* New York: WW Norton.

Schwartz, Peter, and Doug Randall. 2003. *An Abrupt Climate Change Scenario and Its Implications for United States National Security.* Available at: http://www.edf.org/documents/3566_AbruptClimateChange.pdf.

Scott, Shirley. 2008. "Securitizing Climate Change: International Legal Implications and Obstacles." *Cambridge Review of International Affairs* 21 (4): 603–619.

Selby, Jan, Omar S. Dahi, Christiane Fröhlich, and Mike Hulme. 2017. "Climate Change and the Syrian Civil War Revisited." *Political Geography* 60 (September): 232–244.

Smith, Dan, and Janani Vivekananda. 2007. *A Climate of Conflict: The Links between Climate Change, Peace and War.* London: International Alert.

Tickner, J. Ann. 1988. "Hans Morgenthau's Principles of Political Realism: A Feminist. Reformulation." *Millennium: Journal of International Studies* 17 (3): 429–440.

UNEP. 2007. *Sudan: Post-Conflict Environmental Assessment.* Nairobi: UNEP.

Wæver, Ole.1995. "'Securitization and De-Securitization." In *On Security,* edited by Ronnie Lipschutz, 46–87. New York: Columbia University Press.

Walt, Stephen. 1991. "Renaissance of Security Studies." *International Studies Quarterly* 35 (2): 211–239.

Wolf, Aaron T. 1999. "Water Wars" and Water Reality: Conflict and Cooperation Along International Waterways." In *Environmental Change, Adaptation and Human Security,* edited by Steve Lonergan, 251–265. Dordrecht: Kluwer.

3 Global Security Challenges of Climate Change

Halvard Buhaug[1]

Introduction

It can be said that 2018 was a year of meteorological records. Drought, heatwaves, and wildfires struck the northern hemisphere at rates and intensities well outside the range of normal weather, and the year ended as the hottest La Niña year ever recorded and the fourth warmest year overall since recording started in 1850 (WMO 2018). Although attributing any component of these or other specific weather events to anthropogenic climate change remains scientifically challenging, they are consistent with the long-term trend of warming as a result of increased concentration of greenhouse gases in the atmosphere.

The 2018 extremes followed in the wake of another extreme 'event': the European migrant crisis. In 2015, more than one million migrants crossed the Mediterranean Sea in seek of refuge in Europe, and although the numbers have since dropped significantly, they remain well above historical levels. The dominant cause of the upsurge in asylum seekers was a combination of escalating levels of violence in key sending countries (notably Syria, Iraq, Afghanistan, and Nigeria) and increased opportunities for migration by means of human trafficking. Some have also linked the growth in forced displacement with climate change, either directly (Missirian and Schlenker 2017) or indirectly via climatic drivers of social unrest (Kelley et al. 2015).

A third recent series of events that has informed thinking about climate security is the 2011 wave of uprisings across the Middle East and North Africa, commonly referred to as the Arab Spring. The outbreak of protests in Tunisia and Egypt corresponded with a peak in international prices of core food commodities. For Egypt, the world's largest importer of wheat, the increasing cost of food imports eventually led the regime to cut the comprehensive consumer subsidy programme, placing excessive burden on the urban poor (Johnstone and Mazo 2011). A contributing cause of the peaking food prices was major loss of harvest and resulting export bans among major grain producers, due to extreme droughts and heatwaves during the previous growing season (Sternberg 2012).

DOI: 10.4324/9781003001744-4

It is clear that climatic events can have immediate impacts on human security (health, livelihood, food security), but does climate change also constitute a direct threat to peace and societal stability? In this report, I discuss three aspects of relevance to the larger climate security debate: (i) the evident concentration of armed conflict in environmentally fragile regions; (ii) the scientific evidence base for a causal relationship between adverse climatic changes and armed conflict; and (iii) the role of climate-related security threats in a comparative perspective. I end by briefly reflecting on the reverse association, how armed conflict affects climate change vulnerability.

Climate Zones and Conflict Risk

Armed conflicts cluster in space. In the modern, post-Cold War period, large parts of the populated earth have been spared large-scale violent conflict. Other regions have been less privileged. Contemporary civil conflicts are disproportionately located in dry and tropical climates close to the Equator. Since 1950, the rate of civil conflicts has been ten times higher among countries in the dry climate zone than the continental zone (Buhaug, Gleditsch, and Wischnath 2013).

The geographic clustering of contemporary conflicts is clearly discernible in Figure 3.1; central and southern parts of Africa as well as concentrated parts of Asia and Latin America have high to very high densities of conflict events. In contrast, North America, Oceania, and northern Eurasia have largely escaped organised, deadly political violence. In terms of lethality of conflict, some clusters are vastly more destructive than others. In 2017, the wars in Afghanistan, Iraq, and Syria accounted for more than three-quarters of all recorded civil conflict-related battle deaths, according to the Uppsala Conflict Data Program, UCDP (Pettersson and Eck 2018).

One obvious explanation for the clustering of civil conflicts in certain climates is that it reflects—and is a product of—a similar spatial pattern of key risk factors (Buhaug and Gleditsch 2008). Country-specific factors that are robustly associated with increased conflict risk include low level of economic development, fragile and non-democratic governance, natural resource wealth, and a large population (Hegre and Sambanis 2006). None of these features is distributed completely by random. For example, virtually every country in the lower latitudes is classified as a 'developing economy' by the International Monetary Fund, and the poorest countries are also more populous and less politically stable on average than other states. Similarly, the 'bottom billion' controversially identified by Paul Collier a decade ago (Collier 2007; see also Carr-Hill 2013)—inhabitants of countries that are not only poor, but also stagnating or declining—are mostly located around the Equator and particularly in sub-Saharan Africa. Some also see the long-term variation in trajectories of growth and development across regions as a result of local climatic and environmental conditions (e.g., Acemoglu and Robinson 2012; Gallup, Sachs, and Mellinger 1999; Nordhaus 2006).

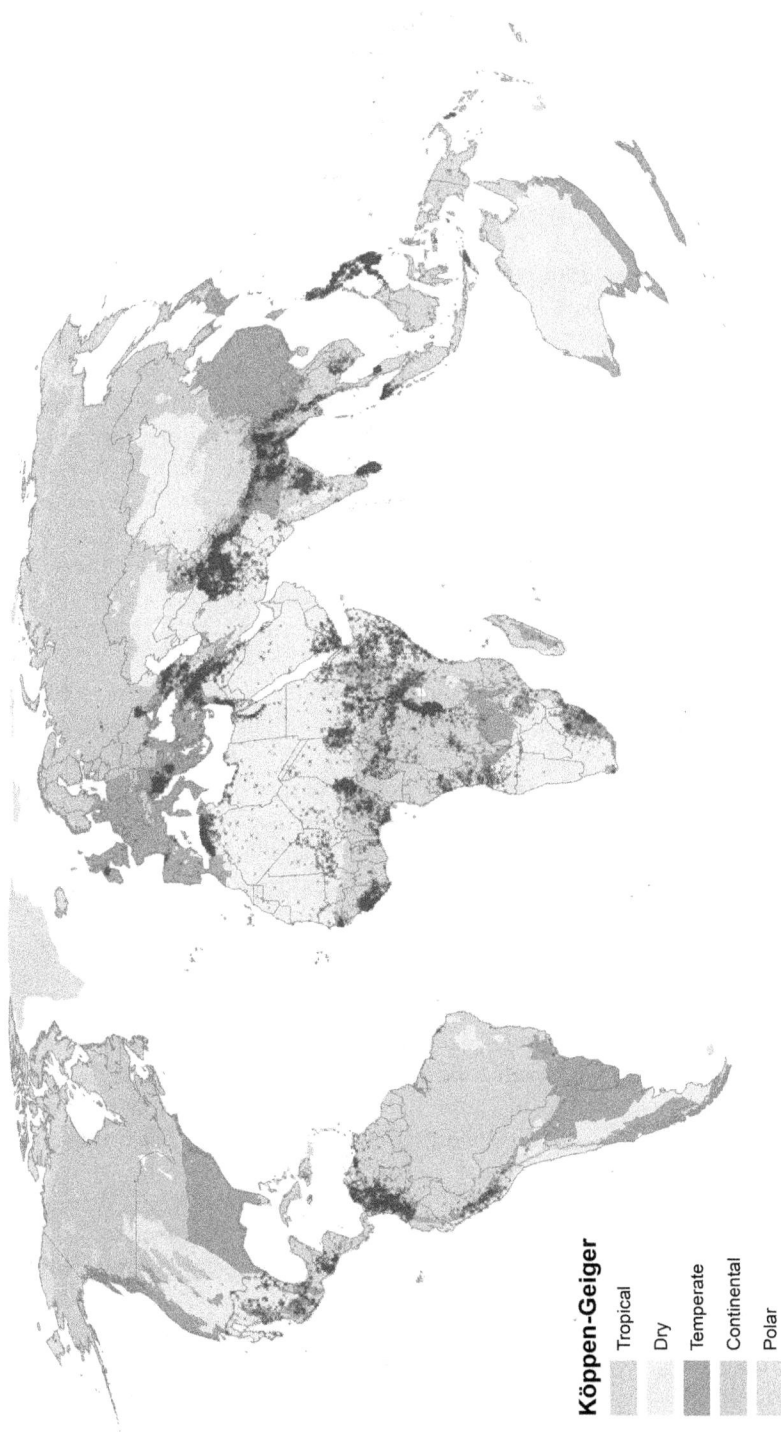

Köppen-Geiger

Tropical

Dry

Temperate

Continental

Polar

Figure 3.1 Location of deadly armed conflict events, 1989–2017[2]

A complementary explanation for the distinct conflict clusters may be that they are results of spillovers from violence in nearby countries. Such a 'bad neighborhood' effect may materialise as a consequence of refugee flows (Salehyan 2008), transnational ethnic linkages (Cederman, Girardin, and Gleditsch 2009), external military intervention (Kathman 2010) or porous borders. We also know that conflict begets conflict through human and material destruction, essentially pushing some countries and regions into a 'conflict trap' (Collier et al. 2003; Gates et al. 2012).

But if conflicts cluster in certain climates, could climate itself be a cause of conflict? No study to date has investigated the relationship between climate, geography, and armed conflict in a rigorous and systematic manner. In fact, neither 'climate' nor 'temperature' features in the index of prominent modern surveys of the causes of war, such as Blainey (1988), Gat (2008) or Suganami (1996). However, the empirical relationship between geography and economic development has been subject to considerable scientific scrutiny. The widely cited study by Gallup, Sachs, and Mellinger (1999) notes a pronounced geographical clustering of countries with low GDP per capita in the tropics. The disparity in overall economic productivity between non-tropical and tropical countries (3.3:1) is even greater for agricultural productivity (8.8:1). The geographical limitations of agriculture in the tropical zone apply equally to humid and arid tropics.

Without subscribing to environmental determinism, it seems fair to assume that armed conflicts cluster in regions with more challenging environmental conditions partly because these conditions have contributed to hindering long-term societal development and facilitated colonial exploitation, and partly because the combination of marginal environments and low socioeconomic development make local social systems more vulnerable to climatic extremes.

The next section briefly reviews the main proposed causal pathways between climatic changes and conflict risk.

Climatic Changes and Conflict Risk

The spatial association between climate zones and conflict prevalence displayed in Figure 3.1 is remarkable, but the aggregate temporal trends in global average temperature and armed conflict are less obviously connected. During the last three decades, when the warming trajectory has become particularly notable, the global number of battle-related deaths from armed conflict has stabilised at the lowest level observed since the Second World War (Figure 3.2). If anything, it would seem that warming is associated with an overall decline in conflict casualties. Of course, such a plot cannot be used to establish causal relationships (or lack thereof), and attempts to make conflict projections based on simple extrapolation of such trends should be treated with much care, but if anything, the figure implies that conflict-affected countries have become better at curbing the level of violence. Could climate change reverse this trend?

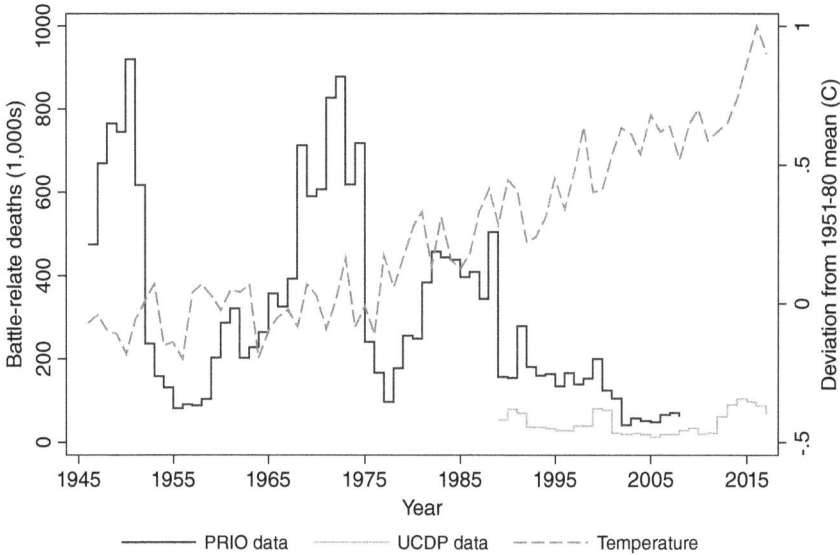

Figure 3.2 Trends in global warming and armed conflict severity, 1946–2017[3]

In 2007, three related events jointly provided a strong impetus for increased investments in scientific research on the climate–security nexus. The first was the release of the UN Intergovernmental Panel on Climate Change (IPCC) Fourth Assessment Report (AR4), which provided the most comprehensive assessment of the state of the art to date on anthropogenic climate change and its likely consequences. The Working Group II contribution to AR4, on Impacts, Adaptation, and Vulnerability, did not explicitly and rigorously consider armed conflict risk (in line with previous reports but unlike the most recent AR5, see Adger et al. 2014). Instead it contained scattered statements about possible negative impacts on societal stability and peace, substantiated primarily by grey literature (Gleditsch and Nordås 2014), effectively revealing a glaring need for more peer-reviewed evidence on the subject.

Second, in April 2007, the UN Security Council held its first ever debate on climate change as a security threat on the initiative of the UK. In her introductory remarks to the debate, then UK Foreign Secretary Margaret Beckett made the connection between climate change and conflict crystal clear: "What makes wars start? Fights over water. Changing patterns of rainfall. Fights over food production, land use" (Lederer 2007). The scientific foundation for this observation remained unclear, however.

The third important event during 2007 was the Norwegian Nobel Committee's decision to give the Nobel Peace Prize to the IPCC and former US Vice-President Al Gore for their work to raise awareness about man-made climate change. As stated in the Nobel Committee's press release announcing the award,

[e]xtensive climate changes may alter and threaten the living conditions of much of mankind. They may induce large-scale migration and lead to greater competition for the earth's resources. Such changes will place particularly heavy burdens on the world's most vulnerable countries. There may be increased danger of violent conflicts and wars, within and between states.

The Nobel Peace Prize was a final reminder of the urgency of the issue and served as a manifestation that the climate security nexus had moved high on the international political agenda and that the debate had run ahead of the scientific evidence base.

Is There a Direct Relationship?

In the decade that has passed since 2007, the collection of academic peer-reviewed studies on climate change and violent conflict has expanded rapidly. In addition to individual studies published in top general science journals that have generated considerable media attention (e.g., Buhaug 2010; Burke et al. 2009; Hsiang and Burke 2013; O'Loughlin, Linke, and Witmer 2014), the research community has produced special issues dedicated to climate and security in leading disciplinary journals, such as the *Journal of Peace Research* (2012; 2021), *Political Geography* (2014), *Geopolitics* (2014), and *Current Climate Change Report* (2017). Overall, this body of research has provided limited evidence that violent conflict is related to climatic changes in a direct and general manner (see Adger et al. 2014 for a comprehensive assessment).

Acknowledging that climatic changes and events are unlikely to influence peace and stability directly and in a sweeping manner, researchers have increasingly shifted their attention towards more plausible indirect pathways and sought to identify conditions under which a climate effect on conflict can be observed. Unsurprisingly, much of this research has focused on Africa, or limited areas within the continent, because of an anticipation that developing countries are especially vulnerable to the impacts of extreme weather events and environmental degradation.[4] Three broad pathways are generally seen as particularly plausible.

Indirect Pathway I: Producer Shocks

The first links climatic changes with political instability and conflict via macroeconomic contraction. In most developing countries, agriculture remains the dominant economic sector for employment and income. Where water availability is scarce or irrigation systems are limited, loss of rainfall can have dramatic impacts on the local economy. Heat-induced stress and destruction of crops due to flooding and saltwater intrusion are other common ways in which extreme weather exerts a negative impact on productivity. Early investigations into this pathway suggested that rainfall patterns indeed are strongly linked with civil conflict risk (Miguel, Satyanath, and Sergenti 2004) although more recent research provide a more mixed impression (Buhaug et al. 2015; Ciccone 2011; Koubi et al. 2012; von Uexkull et al. 2016). However, much of this research is too aggregated to reveal how climatic changes affect household- and community-level economic security, and more research is needed to better understand the observed variation in environmental sensitivity among agricultural economies.

Indirect Pathway II: Consumer Shocks

A second commonly proposed causal pathway views conflict as a possible outcome of weather-induced food price shocks. So-called 'food riots' are not a new phenomenon, but they received increased scholarly attention after the Arab Spring uprisings of 2011, which broke out in the wake of rapidly increasing food prices. For Egypt, the world's largest importer of wheat, the doubling of the international price of wheat meant that the government could no longer sustain its expansive domestic price subsidies, resulting in a tripling of the price of bread in local markets that stirred widespread protest (Sternberg 2012). A number of studies have since uncovered statistically significant correlations between increasing food prices and unrest risk (see Rudolfsen 2020 for a review). The effect appears strongest for low-intensive social events such as demonstrations and riots, most of which never escalate to the level of armed conflict.

While the link between food prices and social unrest is important in its own right, the role of weather and, in the longer term, climate change, in influencing food-related conflict remains less well understood. Historically, climate variability has a modest influence on fluctuations in the international price of food commodities, which instead are driven largely by increasing transportation (oil) and fertiliser costs, global financial downturns, excessive hoarding and market speculation, and more recently, demand for biofuel production (Tadesse et al. 2014). For food markets that rely mostly on local products, the effect of weather shocks will be more pronounced. However, even if conflict may break out during peaking food prices, the protesters are likely to be motivated by other political and economic issues (Sneyd, Legwegoh, and Fraser 2013), implying that stabilising food prices are unlikely to solve underlying grievances in society.

Indirect Pathway III: Forced Migration

A third possible pathway between climatic changes and armed conflict involves forced migration. Extreme weather events and degrading environmental conditions may force people on the margins to relocate and compete for natural resources, public goods, and employment opportunities with host populations, thus escalating latent social conflict and intercommunal animosity. This pathway has been proposed as a contributing cause of the Syrian civil war (Kelley et al. 2015), although the role of climate change in this case is disputed (e.g., Selby et al. 2017).

Due to challenges with ascribing human mobility to specific drivers (see Black et al. 2011 for an insightful discussion) and, consequently, limited data on 'environmental migration,' this mechanism has not been subject to the same level of replicable, generalisable research as the other pathways (though see Reuveny 2007). In a recent review of the literature, Brzoska and Fröhlich (2016, 191) conclude that there is "limited evidence both for the proposition that climate change will lead to major population movements as well as that modern migration movements generally trigger violent conflict."

Research that focuses specifically on refugee flows—for which there is good data—suggests an increased risk of conflict diffusion (Fisk 2014; Salehyan and Gleditsch 2006), although it is not obvious that individuals seeking new economic opportunities as a consequence of climate-induced livelihood loss will have the same destabilising effect as the arrival of traumatised refugees, some of whom may be armed and some of whom may have an aspiration to mobilise new warriors and continue the struggle. Gaining better insight into how environmental and climatic factors interact with other drivers in shaping human mobility (how, when, where) and its knock-on security impacts is a key priority for future research.

It should be mentioned that almost all relevant empirical studies of the three pathways discussed here focus on climate variability, or short-term climatic changes, typically ranging from specific extreme weather events to yearly deviations from long-term mean conditions. Owing to the difficulty of disentangling unique effects of long-term processes and complex systems, social responses to gradual but permanent environmental changes remain poorly understood.

Overall, the more nuanced approach adopted by recent scholarship, investigating the possibility of indirect and conditional climate effects on conflict risk, have not succeeded in uncovering a powerful and 'statistically significant' effect on conflict risk. Results seem to be stronger and more robust for less severe forms of conflict (communal violence, urban riots) than for large-scale civil war, and there is also more consistent evidence that weather fluctuations affect conflict dynamics (severity, duration) than outbreak of conflict in previously stable societies. However, this academic field is still in its infancy, and many research questions remain to be studied or should be revisited using more appropriate analytical tools, and emerging consensus around specific links (e.g., between rising food prices and urban riots) needs to be validated through in-depth case analysis (see Theisen 2017 for a recent review of the scientific literature).

Climatic Drivers in Comparative Perspective

The limited weight of evidence for a causal effect of climatic changes on contemporary violent conflicts should not be used as justification for dismissing the relevance of climate change for societal security altogether. First of all, available research is generally unable to shed light on more complex, long-term relationships between nature and society, which cannot be detected using conventional statistical methods. Second, climatic changes may exhibit a non-linear influence on social systems, permitting coping and adaptation only up to a certain point, beyond which a transition in behaviour may occur—analogous to the final straw that broke the camel's back. As the atmosphere heats up, it will increasingly generate weather phenomena beyond the range of experienced conditions, potentially resulting in cascading and self-reinforcing impacts on ecosystems and societies (see Steffen et al. 2018).

On the other hand, social systems are remarkably inert, and just like today's weather usually constitutes a good prediction of what to expect tomorrow, major sites of armed conflict and their causes are unlikely to change dramatically over time. The best available scientific evidence suggests that climatic impacts on conflict risk to date has been modest, compared to factors such as intergroup inequalities and discrimination, political corruption, weak rule of law, a stagnating economy, a large population, conflictual history, and violent neighbourhood (Mach et al. 2019). Using the past half century as a guide, this suggests that political factors are likely to remain dominant causes of armed conflicts also in the foreseeable future.

Not everybody would agree with this assessment, however. Indeed, there is a tendency among think-tanks and NGO communities to shy away from discussing climate-driven security threats in relative terms. Often, this occurs tacitly, where reports describe how climatic changes may amplify extant risks and outline ways in which climate-driven stressors could translate into various security threats without discussing the role of non-climatic drivers, let alone considering the relative significance of different risk factors. Others are more explicit in their rejection of a comparative approach. Indeed, one of the pioneers in environmental security research admits that he tries to "avoid entangling [himself] in the metaphysical debate about the relative importance of causes," due to the intractable task of separating between causal factors (Homer-Dixon 1999, 7). Likewise, Werrell and Femia (2015) argue that "we need to move away from [...] 'ranking' threats to national security" because of "the complex way in which climate change affects the broader security landscape." I disagree.

The complexity of nature-society interactions, playing out across various temporal and spatial scales and involving feedback loops, requires careful analytical treatment and demands sober conclusions, but it should not prevent us from seeking to gauge the relative importance of contributing factors. As scientists, we have an obligation to offer the best and most rigorous evidence-based insights possible in order to inform the best possible policy. Avoiding efforts to estimate the relative importance of various conflict drivers is not compatible with that ambition, and failing to identify the main causes of social discontent and violent conflict across contexts will hamper the formation of effective peacebuilding policies. Focusing only on climate-related security threats and formulating policy advice that ignores non-climatic security threats not only risks diverting funds and attention away from where they are most effective, it could potentially lead to counter-productive policies (e.g., Hasegawa et al. 2018).

Concluding Remarks

Extreme weather events pose real threats to human security and wellbeing, and climate change is likely to make things worse, especially among societies with people presently living on the margins and lacking necessary skills and resources to cope on their own. So far, however, there is little evidence that climatic changes are an important direct cause of armed conflict, and this chapter has argued that the dominant causes of violent conflict in the years to come are likely to remain political in nature, related to issues such as equality and representation, rule of law, minority protection, and economic wellbeing. Climate change may affect some of these drivers, notably those tied to agricultural production and livelihood security, and hamper development more generally, but heatwaves, crop failures, and weather-induced material destruction are unlikely to result in violent conflict in the absence of other prevailing conflict-promoting conditions. Ongoing armed conflicts, from Afghanistan to Yemen, are fundamentally political contests that require political solutions.

While science is unclear with regard to the true impact of climatic changes on armed conflict, the reverse association, from conflict to climate impacts probably cannot be overstated. Armed conflict is development in reverse (Collier et al. 2003; Gates et al. 2012) and "the biggest threat to human development" (The Millennium Development Goals Report 2015, 8). Accordingly, the single most important strategy to improve local climate resilience and adaptive capacity in conflict-affected societies is to secure a lasting end to fighting, which is necessary in order to attract long-term planning and investments.

Over the past decades, the world has experienced remarkable progress on central human development indicators, for example: a reduction by more than half in the number of people living in extreme poverty since 1990; a reduction by almost half in the number of undernourished people; a reduction by more than half in the under-five mortality rate; a reduction by more than half in the mortality rate of malaria; and considerable improvement in youth literacy rate. As evidenced in Figure 3.2, the world is also becoming less violent, despite recent and horrible setbacks in Syria, Yemen, Northern Nigeria and elsewhere.

These positive developments should not be an excuse for inaction, but it is important not to ignore success stories. While climate change may be the greatest challenge yet to face mankind, we are also better placed than ever before to overcome this challenge. The international community, spearheaded by dominant actors such as the United Nations and the European Union, should increase its investments in finding lasting solutions to ongoing conflicts and avoiding new ones from breaking out. The best way to minimise the security threat imposed by future climate change is to address, and resolve, the dominant causes of contemporary wars.

Notes

1 This work is supported by the European Research Council through grant no. 648291.
2 Data source: UCDP Georeferenced Event Dataset (Sundberg and Melander 2013). The map includes all deadly battle events in civil conflicts, non-state conflicts, and one-sided violence since 1989 (dots; $N = 143,617$ events), imposed on the Köppen-Geiger climate zone classification. Due to the overwhelming nature of the ongoing Syrian civil war, the geocoding of that conflict was not complete at the time of analysis and thus excluded from the map.
3 The figure shows 'best estimate' of global yearly number of battle-related deaths from armed conflict, 1946–2008, from PRIO (Lacina and Gleditsch 2005) (dark gray solid line); global yearly 'best estimate' number of battle-related deaths from armed conflict, 1989–2017, from UCDP (Pettersson and Eck 2018) (light gray solid line); and global yearly average combined land-surface air and sea-surface water temperature anomalies, compared to the global 1951–1980 mean, from NASA (dashed line). The PRIO and UCDP battle deaths datasets use similar definitions of armed conflict, but UCDP adopts a more stringent methodology for classifying information as credible, resulting in more conservative estimates.
4 While it often makes sense to zoom in on those cases considered especially likely to host climate-affected conflicts, this sampling strategy is not without limitations, see, for example, Adams et al. (2018), Gleditsch (1998) and Hendrix (2017).

References

Acemoglu, Daron, and James A. Robinson. 2012. *Why Nations Fail: The Origins of Power, Prosperity, and Poverty.* 1st edition. New York: Currency.
Adams, Courtland, Tobias Ide, Jon Barnett, and Adrien Detges. 2018. "Sampling Bias in Climate–Conflict Research." *Nature Climate Change* 8 (3): 200–203.
Adger W. Neil, Juan M. Pulhin, Jon Barnett, Geoffrey D. Dabelko, Grete K. Hovelsrud, Marc Levy, Ürsula Oswald Spring, et al. IPCC 2014. "Human Security". In *Climate Change 2014. Impacts, Adaptation, and Vulnerability. Part A: Global and Sectoral Aspects.* Contribution of Working Group II to the Fifth Assessment Report of the Intergovernmental Panel on Climate Change, edited by C.B. Field, V.R. Barros, D.J. Dokken, K.J. Mach, M.D. Mastrandrea, T.E. Bilir, M. Chatterjee, K.L. Ebi, Y.O. Estrada, R.C. Genova, B. Girma, E.S. Kissel, A.N. Levy, S. MacCracken, P.R. Mastrandrea, and L.L. White, 755–791. Cambridge: Cambridge University Press.
Black, Richard, W. Neil Adger, Nigel W. Arnell, Stefan Dercon, Andrew Geddes, and David Thomas. 2011. "The Effect of Environmental Change on Human Migration." *Global Environmental Change* 21: S3–11.
Blainey, Geoffrey. 1988. *The Causes of War.* 3rd edition. New York: Free Press.
Brzoska, Michael, and Christiane Fröhlich. 2016. "Climate Change, Migration and Violent Conflict: Vulnerabilities, Pathways and Adaptation Strategies." *Migration and Development* 5 (2): 190–210.
Buhaug, Halvard. 2010. "Climate Not to Blame for African Civil Wars." *Proceedings of the National Academy of Sciences* 107 (38): 16477–16482.
Buhaug, Halvard, Tor A Benjaminsen, Espen Sjaastad, and Ole Magnus Theisen. 2015. "Climate Variability, Food Production Shocks, and Violent Conflict in Sub-Saharan Africa." *Environmental Research Letters* 10 (12): 125015.

Buhaug, Halvard, and Kristian Skrede Gleditsch. 2008. "Contagion or Confusion? Why Conflicts Cluster in Space." *International Studies Quarterly* 52 (2): 215–233.

Buhaug, Halvard, Nils Petter Gleditsch, and Gerdis Wischnath. 2013. "War(m) Zones: Climate, Development, and Civil Conflict". Unpublished manuscript, PRIO.

Burke, Marshall, Edward Miguel, Shanker Satyanath, John A. Dykema, and David B. Lobell. 2009. "Warming Increases the Risk of Civil War in Africa." *Proceedings of the National Academy of Sciences* 106 (49): 20670–20674.

Carr-Hill, Roy. 2013. "Missing Millions and Measuring Development Progress." *World Development* 46 (June): 30–44.

Cederman, Lars-Erik, Luc Girardin, and Kristian Skrede Gleditsch. 2009. "Ethnonationalist Triads: Assessing the Influence of Kin Groups on Civil Wars." *World Politics* 61 (3): 403–437.

Ciccone, Antonio. 2011. "Economic Shocks and Civil Conflict: A Comment." *American Economic Journal: Applied Economics* 3 (4): 215–227.

Collier, Paul. 2007. *The Bottom Billion: Why the Poorest Countries Are Failing and What Can Be Done About It.* 1st edition. Oxford: Oxford University Press.

Collier, Paul, V.L. Elliott, Håvard Hegre, Marta Reynal-Querol, and Nicholas Sambanis. 2003. *Breaking the Conflict Trap: Civil War and Development Policy.* Washington, D.C.: World Bank and Oxford University Press.

Fisk, Kerstin. 2014. "Refugee Geography and the Diffusion of Armed Conflict in Africa." *Civil Wars* 16 (3): 255–275.

Gallup, John Luke, Jeffrey D. Sachs, and Andrew D. Mellinger. 1999. "Geography and Economic Development." *International Regional Science Review* 22 (2): 179–232.

Gat, Azar. 2008. *War in Human Civilization.* Oxford: Oxford University Press.

Gates, Scott, Håvard Hegre, Håvard Mokleiv Nygård, and Håvard Strand. 2012. "Development Consequences of Armed Conflict." *World Development* 40 (9): 1713–1722.

Gleditsch, Nils Petter. 1998. "Armed Conflict and The Environment: A Critique of the Literature." *Journal of Peace Research* 35 (3): 381–400.

Gleditsch, Nils Petter, and Ragnhild Nordås. 2014. "Conflicting Messages? The IPCC on Conflict and Human Security." *Special Issue: Climate Change and Conflict* 43 (November): 82–90.

Hasegawa, Tomoko, Shinichiro Fujimori, Petr Havlík, Hugo Valin, Benjamin Leon Bodirsky, Jonathan C. Doelman, Thomas Fellmann, et al. 2018. "Risk of Increased Food Insecurity under Stringent Global Climate Change Mitigation Policy." *Nature Climate Change* 8 (8): 699–703.

Hegre, Håvard, and Nicholas Sambanis. 2006. "Sensitivity Analysis of Empirical Results on Civil War Onset." *Journal of Conflict Resolution* 50 (4): 508–535.

Hendrix, Cullen S. 2017. "The Streetlight Effect in Climate Change Research on Africa." *Global Environmental Change* 43 (March): 137–147.

Homer-Dixon, Thomas F. 1999. *Environment, Scarcity, and Violence.* Ewing: Princeton University Press.

Hsiang, Solomon M., and Marshall Burke. 2013. "Climate, Conflict, and Social Stability: What Does the Evidence Say?" *Climatic Change* 123 (1): 39–55.

Johnstone, Sarah, and Jeffrey Mazo. 2011. "Global Warming and the Arab Spring." *Survival* 53 (2): 11–17.

Kathman, Jacob D. 2010. "Civil War Contagion and Neighboring Interventions." *International Studies Quarterly* 54 (4): 989–1012.

Kelley, Colin P., Shahrzad Mohtadi, Mark A. Cane, Richard Seager, and Yochanan Kushnir. 2015. "Climate Change in the Fertile Crescent and Implications of the Recent Syrian Drought." *Proceedings of the National Academy of Sciences* 112 (11): 3241–3246.

Koubi, Vally, Thomas Bernauer, Anna Kalbhenn, and Gabriele Spilker. 2012. "Climate Variability, Economic Growth, and Civil Conflict." *Journal of Peace Research* 49 (1): 113–127.

Lacina, Bethany, and Nils Petter Gleditsch. 2005. "Monitoring Trends in Global Combat: A New Dataset of Battle Deaths." *European Journal of Population/Revue Européenne de Démographie* 21 (2–3): 145–166.

Lederer, Edith M. 2007. "Security Council Tackles Climate Change." The Associated Press, 18 April. http://www.washingtonpost.com/wp-dyn/content/article/2007/04/18/AR2007041800219_pf.html.

Mach, Katharine J., Caroline M. Kraan, W. Neil Adger, Halvard Buhaug, Marshall Burke, James D. Fearon, Chris B. Field, Cullen S. Hendrix, Jean-Francois Maystadt, John O'Loughlin, Philip Roessler, Jürgen Scheffran, Kenneth A. Schultz, and Nina von Uexkull. 2019. "Climate as a Risk Factor for Armed Conflict." *Nature* 571: 193–197.

Miguel, Edward, Shanker Satyanath, and Ernest Sergenti. 2004. "Economic Shocks and Civil Conflict: An Instrumental Variables Approach." *The Journal of Political Economy* 112 (4): 725–753.

Missirian, Anouch, and Wolfram Schlenker. 2017. "Asylum Applications Respond to Temperature Fluctuations." *Science* 358 (6370): 1610.

Nordhaus, William D. 2006. "Geography and Macroeconomics: New Data and New Findings." *Proceedings of the National Academy of Sciences of the United States of America* 103 (10): 3510–3517.

O'Loughlin, John, Andrew M. Linke, and Frank D. W. Witmer. 2014. "Effects of Temperature and Precipitation Variability on the Risk of Violence in Sub-Saharan Africa, 1980–2012." *Proceedings of the National Academy of Sciences* 111 (47): 16712–16717.

Pettersson, Thérése, and Kristine Eck. 2018. "Organized Violence, 1989–2017." *Journal of Peace Research* 55 (4): 535–547.

Reuveny, Rafael. 2007. "Climate Change-Induced Migration and Violent Conflict." *Political Geography* 26 (6): 656–673.

Rudolfsen, Ida. 2020. "Food Insecurity and Domestic Instability: A Review of the Literature." *Terrorism and Political Violence* 32 (5): 921–948.

Salehyan, Idean. 2008. "The Externalities of Civil Strife: Refugees as a Source of International Conflict." *American Journal of Political Science* 52 (4): 787–801.

Salehyan, Idean, and Kristian Skrede Gleditsch. 2006. "Refugees and the Spread of Civil War." *International Organization* 60 (2): 335–366.

Selby, Jan, Omar S. Dahi, Christiane Fröhlich, and Mike Hulme. 2017. "Climate Change and the Syrian Civil War Revisited." *Political Geography* 60 (September): 232–244.

Sneyd, Lauren Q., Alexander Legwegoh, and Evan D.G. Fraser. 2013. "Food Riots: Media Perspectives on the Causes of Food Protest in Africa." *Food Security* 5 (4): 485–497.

Steffen, Will, Johan Rockström, Katherine Richardson, Timothy M. Lenton, Carl Folke, Diana Liverman, Colin P. Summerhayes, et al. 2018. "Trajectories of the Earth System in the Anthropocene." *Proceedings of the National Academy of Sciences*, in press.

Sternberg, Troy. 2012. "Chinese Drought, Bread and the Arab Spring." *Applied Geography* 34 (May): 519–524.

Suganami, Hidemi. 1996. *On the Causes of War*. Oxford: Oxford University Press.

Sundberg, Ralph, and Erik Melander. 2013. "Introducing the UCDP Georeferenced Event Dataset." *Journal of Peace Research* 50 (4): 523–532.

Tadesse, Getaw, Bernardina Algieri, Matthias Kalkuhl, and Joachim von Braun. 2014. "Drivers and Triggers of International Food Price Spikes and Volatility." *Food Policy* 47 (August): 117–128.

The Millennium Development Goals Report 2015. United Nations. https://resource-centre.savethechildren.net/library/millennium-development-goals-report-2015.

Theisen, Ole Magnus. 2017. "Climate Change and Violence: Insights from Political Science." *Current Climate Change Reports* 3 (4): 210–221.

von Uexkull, Nina, Mihai Croicu, Hanne Fjelde, and Halvard Buhaug. 2016. "Civil Conflict Sensitivity to Growing-Season Drought." *Proceedings of the National Academy of Sciences* 113 (44): 12391–12396.

Werrell, Caitlin, and Francesco Femia. 2015. "'Is Climate Change the Biggest Security threat?' Is Still a Bad Question." Center for Climate and Security. https://climateandsecurity.org/2015/11/16/is-climate-change-the-biggest-security-threat-is-still-a-bad-question/.

WMO 2019. *WMO Statement on the State of the Global Climate in 2018*. Geneva: World Meteorological Organization.

Part II

Climate Change and Conflict in the Pacific

4 Climate Change, Its Social Effects and Conflicts in the Pacific

Volker Boege

Introduction

It is generally acknowledged that islands and coastal regions will be particularly severely impacted by climate change (Nurse et al. 2014). This holds true first and foremost for the Pacific Island countries (PIC). Many PIC are particularly vulnerable due to their extreme exposure and their rather constrained options for adaptation. Sea level rise, the increased frequency and severity of extreme weather events such as tropical cyclones and storm surges, floods and droughts, coastal erosion, salt water intrusion and salinisation and other natural hazards, challenge island economies and habitats as well as the livelihoods of people in the region. Food, land and water security are under pressure, and a broad spectrum of newly arising economic, social and cultural problems can be attributed to the effects of climate change.

The economic, environmental, social, cultural and other effects of climate change are conflict-prone. In the Pacific region, conflicts over land and scarce natural resources, conflicts due to climate change-induced migration, or conflicts arising from poor environmental governance or poorly designed and implemented climate change adaptation and mitigation responses are cases in point. Consequently, the regional organisation for the Pacific, the Pacific Islands Forum (PIF), in its Regional Security Declaration of 2018 (the Boe Declaration) stated that "climate change remains the single greatest threat to the livelihoods, security and wellbeing of the peoples of the Pacific" (Pacific Islands Forum 2018). In the Declaration, the PIF refers to an "expanded concept of security" which includes "environmental and resource security" and has a focus on "emerging security challenges" (ibid.).

Against the background of these concerns, it is worthwhile to explore the nexus between climate change, its social effects, conflict and security in the Pacific in more detail. This chapter is to make an initial contribution to such an endeavour, in a rather cursory way, giving an overview of actual and potential linkages between climate change and conflict in the Pacific.

The chapter is structured as follows. First, the environmental and social effects of climate change on Pacific Island countries are sketched very briefly. Second, the climate change–conflict nexus will be discussed. Third, how

DOI: 10.4324/9781003001744-6

governance and policies (can or should) intervene in order to prevent climate change–induced conflict will be explored; it will be argued that governance is the decisive link in the climate change–conflict nexus. Flowing from that, options for conflict-sensitive approaches to the challenges of climate change policies in Oceania will be presented, highlighting the importance of non-state customary actors and institutions and of indigenous non-Western approaches to climate change adaptation as well as to conflict transformation and peacebuilding. In conclusion, a case will be built for an integrated and holistic multi-layered governance framework which encompasses actors and institutions at various scales and from different societal spheres.

The Environmental and Social Effects of Climate Change in Oceania

If the small islands states of the Pacific are on the radar of politics and the wider public outside of the region these days, then it is first and foremost in the context of climate change. The sinking islands of the Pacific have become a symbol for the severe unprecedented consequences of man-made global warming, foreshadowing climate change-related environmental and social developments that will affect other parts of the world sooner rather than later.

The Pacific region is extremely diverse in many respects: geographically, economically, socially, politically, linguistically and culturally. In today's international political system, the region is divided into 'nation'-states, most of them very small by international standards, many of them comprising dozens of islands. The 22 countries and self-governing territories altogether have a population of approximately 10 million people who inhabit about 300 islands (out of around 7500 islands altogether). Of the 32 million square kilometres of the region, 98 percent is water. Of the remaining 2 percent of land mass Papua New Guinea (PNG) alone comprises about 95 percent. With its approximately seven million inhabitants, PNG by far has the biggest population. None of the other PIC has a population of over one million. The region has the greatest concentration of micro-states (states with less than half a million inhabitants) worldwide.

Apart from the independent states and self-governing territories there are several political entities with a colonial or quasi-colonial status. Decolonisation in the region occurred relatively late, between 1962 (independence of Samoa) and 1980 (independence of Vanuatu). The residues of colonialism strongly reverberate in the region. French Polynesia and Wallis and Futuna are overseas French territories, and so is New Caledonia, albeit with a special political status and a series of referenda on the option of political independence.[1] Niue, the Cook Islands and Tokelau have special relationships with New Zealand (in 'free association' with New Zealand). Other self-governing territories are legally linked to the USA: the territories of the Northern Mariana Islands, the Federated States of Micronesia, the Marshall Islands, Palau and American Samoa. Finally, some islands or territories in the Pacific region are part of non-region states: Rapa Nui (Easter Island) is part of Chile, Hawaii is part of the USA, the Torres Straits Islands are part of Australia, and (West) Papua which occupies the western half of the island of New Guinea is part of Indonesia; this status, however, is strongly contested by an indigenous movement for self-determination.[2]

All the islands in the region are subject to the environmental effects of climate change, in particular sea level rise, an increased frequency and intensity of extreme weather events such as tropical cyclones and storm surges, increasing air and sea surface temperatures, and changing rainfall patterns, including protracted droughts (Nurse/IPCC 2014, 1616).[3]

Sea-level rise and associated submersion, storm surges, salt water intrusion, salinisation, erosion and other coastal hazards degrade fresh groundwater resources and reduce land available for agriculture, settlements and infrastructure. Sea surface temperature rise results in increased coral bleaching and reef degradation, leading to a reduction of fish stocks and as a consequence declining fish catch (ibid.). Rising temperatures will also increase the risk of vector-borne diseases such as malaria and dengue fever as well as diarrhoeal diseases, with significant ramifications for health sectors in PIC.

The particularly high level of climate change-related vulnerability[4] of many islands in Oceania is due to their extreme exposure and their rather constrained options for adaptation. This holds particularly true for atoll islands which are of extremely low elevation and often also of rather limited area. The highest point of the Pacific Island country of Tuvalu is 1.50 metres above sea level, for Kiribati it is three metres, and the average island width of Kiribati islands is less than 1000 meters. Atoll countries are particularly vulnerable to sea level rise "because of their high ratio of coastline to land area, relative high population densities and low level of available resources for adaptive measures" (Yamamoto and Esteban 2010, 2). Large islands with high elevations and volcanic high islands are less exposed, but also face severe climate change-induced environmental degradation, particularly along their coastlines.

Given the environmental effects of climate change, PIC are confronted with challenges to land security, livelihood security and habitat security (Campbell 2014, 4–5), which includes water security and food security, as well as health. Land security is under threat due to coastal erosion and inundation, livelihood and habitat security due to reduced quantity and quality of water supplies and loss of food production. These losses affect atoll communities in particular, but also coastal locations, river delta or inland river communities.

Fertile soils are scarce on many islands, and sea water intrusion causes soil salinisation and contamination of freshwater lenses which provide people with water for drinking and agriculture. Most people depend on traditional subsistence agriculture, supplemented by some cash cropping. This is the basis of their way of living. It is under growing pressure as yields from gardens and freshwater supplies decline. As a consequence, increasing numbers of people become more and more dependent on aid from outside.

Of course, people try to adapt to the impacts of climate change. But given the geographical conditions, options for *in situ* technical adaptation[5]—such as planting mangroves in order to reduce coastal erosion, building seawalls in order to contain storm surges and king tides, setting up rainwater tanks to improve fresh water supply—are limited. In many cases they are technically not feasible or too costly, and sometimes they only work as interim measures. Consequently, migration or resettlement to locations that are less exposed might be the better—or even the only—option of long-term sustainable adaptation in certain cases.

Migration can be seen as an alternative to *in situ* adaptation or as another adaptation strategy. Views vary on whether it is an adaptation measure among others (migration as part of an integrated adaptation strategy), or whether it is an adaptation measure of last resort only, once a location has become (almost) uninhabitable. In fact, relocation in

> some extreme circumstances (…) is likely to be the only option left when the life-support systems (land, livelihood, and/or habitat security) of a community's territory fail. In such cases, the migration becomes forced, and the movement may involve whole or large portions of communities.
>
> (Campbell 2014, 7)

Some communities actually have been forced to relocate already, and climate change-induced migration (climate migration)[6] will become a growing concern, given that many islands and even entire small island states are under threat of becoming uninhabitable or even submerged by rising seas.[7]

Climate Change and (Violent) Conflict

Over the last decade, research into the climate change–conflict nexus has gained considerable attention in peace and conflict studies (as well as in security studies). Research and findings have become ever more complex and sophisticated, trying to disentangle the "long and uncertain causal chains from climate change to social consequences like conflict" (Gleditsch, Nordas and Salehyan 2007, 8).[8] Research points to the environmental effects of climate change (e.g. sea level rise), which, in turn, have economic and social effects (e.g. economic decline, loss of livelihood or forced migration), and these effects can lead to violence and violent conflict if certain political and societal conditions prevail, such as fragile statehood, poor governance or deep horizontal or vertical social fragmentation.[9]

Taking migration as an example for a crucial link in the climate–conflict chain, such "causal chains" can go like this: people forced from their homelands due to the environmental and social effects of climate change (e.g. sea level rise, water scarcity, food insecurity) clash with people in recipient regions over scarce natural resources, employment opportunities, cultural differences, etc. (the climate change–migration–conflict chain). Or: climate change leads to environmental degradation which leads to violent conflicts (over land and/ or water), and such violent conflict leads to migration (the climate change–conflict–migration chain) (Reuveny 2007, 660).

In fact, migration is seen as "one of the most plausible links from climate change to conflict" as Nils Petter Gleditsch and colleagues found more than a decade ago (Gleditsch, Nordas and Salehyan 2007, 4). And Dan Smith and Jani Vivekananda, also in 2007, identified migration as a key conflict-relevant risk of climate change (Smith and Vivekananda 2007, 21–22). More recent takes on this topic are the G7-commissioned report 'A new climate for peace' by Adelphi and others from April 2015, which also makes the link between climate change, social disruption, migration and "local and regional instability" (Rüttinger et al. 2015, 3), or a USAID Discussion Paper from November 2016, which addresses "climate, migration, and conflict", presenting an in-depth analysis of the high-profile cases of Darfur and Syria (Null and Rizi 2016).[10]

Other "causal chains" in the climate change–conflict nexus mention soil degradation and desertification, caused by climate change, leading to food insecurity and conflicts over water, pastures and arable land, or degradation of freshwater resources, caused by climate change, leading to conflicts over water.[11] One has to be careful, however, not to oversimplify such causal chains and fall into the trap of naive determinism.

Researchers are in agreement today that there is no direct causal link between climate change and violent conflict, hence talk about 'climate wars' should be avoided. Rather, the term 'climate change-induced violent conflict' seems more appropriate. This term both stresses the significance of climate change for certain conflicts and puts climate change into perspective as one among other causal factors. In other words: climate change and its environmental, social and other effects can be factors in a complex conflict-prone societal constellation and a multi-staged process which can lead to the violent conduct of conflict, or even war. To approach such conflicts from the 'climate change' angle and through the lens of 'climate change and conflict' is valid if it can be hypothesised that climate change and its effects play a dominant role in the conflict constellation and the escalation of conflict; whether this is actually the case can only be explored by thorough case study research. Hence one has to avoid reductionist and deterministic narratives (such as: climate change leads to resource scarcity leads to violent conflict) and instead pay due attention to the entirety of factors which constitute conflict-prone constellations and pathways. In particular, one has to take note of "the importance of institutions and quality of governance" (Buhaug 2015, 271).

A recent assessment of the conflict risks posed by climate change globally comes to the conclusion "that climate fragility risks persist and are worsening" (Vivekananda 2017, 41). And the greatest risks "emerge when the impacts of climate change overburden weak states. Climate change is the ultimate "threat multiplier": it will aggravate already fragile situations and may contribute to social upheaval and even violent conflict" (Ruettinger et al. 2015, 1).

For Oceania, however, the climate change–conflict nexus has not yet been explored explicitly. Violent conflict in Oceania is less prominent a topic than climate change (or migration), mainly because violent conflicts in the region appear as being rather minor in comparison to today's major wars, like in Syria or Afghanistan, and they take place far away from the global power centres.

Indeed, there are no "climate wars" in the Pacific, neither inter-state nor intra-state. What can be found, however, are conflicts in the local context which at times escalate violently, albeit at a relatively low level of intensity, and what can be found is everyday dispersed violence, such as domestic violence against women and children. This everyday violence and these local low-intensity violent conflicts can be often linked to the social effects of climate change, in particular climate change-induced migration and relocation.

A case in point, for example, is the widespread domestic violence in the overcrowded squatter settlements of the few urban centres in PIC. These settlements are also often the sites of violent, sometimes deadly, conflicts between communities from different islands—communities whose members often have left their home islands also due to the effects of climate change.

Causation of course is always complex and path dependent. The deadly clashes between two island communities (from the islands of Tanna and Ambrym) in Port Vila, the capital city of Vanuatu, for example, in 2007 were triggered by sorcery accusations (Boege and Forsyth 2018). It would be misleading, however, to label this conflict as a 'sorcery conflict'. But it would also be misleading to label it a 'climate conflict' on the grounds that members of these communities had migrated to Port Vila due to—inter alia[12]—the effects of climate change on their home islands. Both climate change and sorcery (accusations) played a role in causing this conflict, and so did a host of other factors. 'Climate change-induced migration' can be identified as one element in a complex web of conflict causation, and sorcery (allegations) can be identified as a trigger in a complex process of conflict escalation. The fact that this conflict escalated violently is due to a specific constellation of factors and events which, taken together, 'caused' the violent conduct of conflict in a non-linear complex and emergent manner. For the people directly involved in this conflict (or at least most of the people), sorcery was without doubt its cause. Western academics analysing the conflict will reject this explanation and turn to more 'rational' or 'objective' causes (like climate change-induced migration). In doing so, however, they miss an important point, namely the worldviews, perceptions and motivation of the people on the ground. Hence conflicts like this one have to be grasped as complex and emergent (which also includes the appearance of non-predictable unexpected/contingent phenomena) rather than as 'caused' by climate change (-induced migration)—or sorcery.

Localised violent conflicts do not only occur in the urban settlements of migrants from climate change affected islands, but also on the islands themselves. Here conflicts over the scarce natural resource, land, emerge between people moving from the coast to higher ground and the people already living there. Or, in Kiribati water scarcity has led to conflicts over water between neighbouring communities which felt forced to encroach on each other's land (Foreign Affairs Committee 2010, 102).

Of particular concern is the situation in the Solomon Islands and the Autonomous Region of Bougainville in Papua New Guinea which both experienced large-scale internal violent conflicts not long ago and which still are in post-conflict peacebuilding situations.

In the Solomon Islands, the island and province of Malaita was a conflict hotspot during the civil war of the early 2000s. Currently some outer islands of Malaita province are becoming uninhabitable due to sea-level rise and its effects, and people have started to relocate to mainland Malaita—more or less spontaneously or organised. On occasions, there have been outbursts of sporadic violence. Malaita is the most densely populated island in the Solomon Islands, it is categorised as overpopulated by the government, and land is extremely scarce. Over the last decades, thousands of Malaitans have migrated to other parts of the Solomon Islands, mainly to the capital city Honiara on the island of Guadalcanal, and conflicts between Malaitan settlers and the local population on Guadalcanal had been a major factor in the large-scale violent conflict of the early 2000s.

In the case of Bougainville, planned relocation from the islands of the Carterets atoll to mainland Bougainville has been going on for a couple of years now (see Boege and Rakova in this volume). Although this in the main is a relatively successful exercise, there have been disputes over land which even led to the re-relocation of Carterets islanders back home to their islands from the resettlement site in Tinputz on mainland Bougainville. And people from another Carterets relocation site on Buka island (a major island adjacent to Bougainville Island) reported that there was "a lack of 'unity' with the host community" (Lange 2009, 103), with ongoing conflicts over land use and fishing rights. Relocatees were the target of hostilities from their neighbours who destroyed their houses and food gardens or their produce when they took it to the market or attacked their young people or raped the women (Lange 2009, 104). As a consequence, "many families returned to the Carteret Islands due to difficulties integrating with the host community" (ibid.).

These examples demonstrate that even planned relocation can lead to local conflicts between settlers and recipient communities. And they demonstrate that one has to look beyond state-based violent conflict (interstate or intrastate wars) and to also take into account inter- and intra-group violence in local and everyday contexts. This kind of "intergroup violence below the state level", however, usually remains under the radar of research into the climate change–conflict nexus (and it slips through the grids of large-N studies).[13]

These types of low intensity violence and conflicts may look petty and negligible in comparison to the popular 'climate wars' talk. For the people who are directly affected, however, these 'small' conflicts are extremely serious; for them they can have devastating, even life-threatening or deadly, consequences.

They can be interpreted as conflicts of interest over scarce natural resources (land, water, fish stocks,…) and conflicts of interest over access to public goods and economic opportunities (jobs in the formal economy, access to health and education facilities,…), and some are also identity conflicts (over customs, culture, history, religion, sorcery,…). As Pacific Islands are usually "regions with exclusive identities", they are particularly prone to such identity conflicts (Brzoska and Froehlich 2015, 14–15). Climate change-induced migration is without doubt a contributing factor in the conflicts mentioned. Its specific significance, however, can only be explored on a case-by-case basis, in the overall context of complex and emergent conflictual webs and pathways.

Having said that, violent conflict escalation cannot be ruled out, particularly in a fragile post-conflict environment such as in Bougainville or Solomon Islands, or in situations that are already conflict-prone anyway due to other factors which are of relevance for the interests and identities of (potential) conflict parties.[14] In those situations climate change-induced migration, for example, as well as other effects of climate change, can lead to conflict escalation, particularly in the resettlement areas (be they urban squatter settlements or rural communally owned lands), between newcomers and locals or between different groups of newcomers, particularly under conditions of scarcity and (perceptions of) inequality. Here climate change is a 'threat multiplier'.

Whether such conflicts will actually escalate violently or not, however, is dependent on additional intervening factors and their relations and interactions, in particular on the actions and reactions of affected communities, including the history of relations between them, the (dis)functionality of customary, state and civil society dispute resolution mechanisms, the adaptive capacity of affected communities (including options for long-distance migration or planned relocation), the capabilities and preparedness to use physical force in conflict situations, and not least the stability or fragility of the overall societal and political environment.

Flowing from these observations, it can be posited that whether climate change-induced conflicts escalate violently or not depends primarily on the four following variables:

1 The gravity and urgency of the climate change-induced environmental degradation;
2 The vulnerability and adaptive capacity of affected communities (what are the options to adapt, to change lifestyles or to relocate?);
3 The capacity and willingness to use violence as a means to conduct conflict and to 'solve' problems);[15]
4 The fragility or stability of the societal and political context, or, to put it differently: violent conflict escalation is likely to occur in fragile situations in which state institutions are weak and lack capacity, effectiveness and legitimacy and in which other—non-state customary and/or civil society—avenues for addressing the effects of climate change and for the non-violent conduct of conflict are also absent or insufficient.

Whereas the first point lies beyond the reach of political intervention (at least in the local context), the other points can be addressed by conflict-preventive and conflict-sensitive politics. In other words: whether there will be violent escalation of climate change-induced conflicts or not, in the first place depends on governance.

Governance is a crucial node in the complex network of elements and relations which constitute the specific emergent conflict constellations. Talking about governance has to include, but at the same time also to transcend, issues of weakness and fragility or strength and stability of states. In countries like the PIC, hybridity of political order and governance arrangements have to be taken into account.

Climate Change-Related Governance and Peacebuilding

The starting point for engaging with climate change governance as of major significance for the climate change–conflict nexus is the insight that "(p)olitical, economic, and social contexts are as important to understanding vulnerability as exposure to the physical effects of climate change itself" (Null and Rizi 2016, 5).

In a fragile post-conflict environment (such as in Bougainville or Solomon Islands), or under conditions of state fragility more generally, climate change governance poses particular challenges. PIC with their limited institutional capacities have much more difficulties in dealing with the effects of climate change than stable states (the 'climate-fragility risk' (Rüttinger et al. 2015)). Lack of capacities and ensuing lack of effectiveness diminishes the legitimacy and trustworthiness of state institutions in the eyes of the people on the ground, and lack of legitimacy makes it more difficult for state institutions to effectively implement adaptive measures, for example planned relocation.

In such fragile situations non-state actors can and do play important roles. This does not only include civil society organisations in the Western understanding of the term, but also traditional authorities and institutions from the local customary sphere of societal life. In PIC the resilience of communities and their adaptive capacity very much rests with densely knit customary societal networks of support and reciprocity, with customary authorities and institutions as effective and legitimate governance actors and mechanisms. Hence climate change and its effects are not issues that can be dealt with in the framework of the state and its institutions alone, but local customary non-state, as well as civil society, institutions have to be included.

Traditional authorities—chiefs and elders, tribal leaders, religious authorities, healers, big men and wise women—are of major importance for the organisation of everyday life in Pacific societies. They are in charge of the governance of communities, natural resources and the environment, they regulate resource use and solve disputes (not least disputes over land and other natural resources) according to local custom. They have to be taken into account and included when it comes to the management of the effects of climate change and adaptation policies. Resilience of communities and adaptive capacity very much rest with customary landownership and the associated customary actors and institutions and the indigenous traditional knowledge of which they are custodians (Bryant-Tokalau 2018). They are important for planning, decision-making and implementation of climate change adaptation programmes. "The incorporation of traditional culture (*kastom*), knowledge and local leadership (Chiefs) will ensure that initiatives are appropriate and engage local communities" (Davies 2019, 17).

Communities' adaptive capacity—seen not as a technical issue, but in its political, cultural and social dimensions—to a large extent rests with customary actors and institutions. In particular, they are of major significance for a holistic approach to the 'land' issue with all its aspects, including the 'soft'—cultural, psychological, spiritual—dimensions.[16]

This of course is not to say that governments and state institutions do not matter in climate change governance. They have the power to set framework conditions for climate mitigation and adaptation at the national level, and they provide the link between needs and interests of local populations and the international level, by representing their people in international climate politics and by securing international assistance for climate adaptation measures in their respective countries, either directly (e.g. via the Green Climate Fund) or indirectly via development assistance, which in the PIC increasingly comprises climate change-related programmes and projects.

Finally, the role of the churches as the most important civil society organisations in PIC cannot be overestimated. The vast majority of Pacific Islanders are devout Christians. State institutions in PIC might not reach far beyond the urban centres, but the churches are everywhere on the ground. They can provide valuable leadership in adaptation and the governance of climate change effects more generally.

The network of state and non-state, civil society and customary actors and institutions which are all involved in the governance of communities and societies can be conceptualised as hybrid political order. In hybrid political orders, diverse and competing authority structures, sets of rules and logics of order co-exist, compete, overlap, interact, intertwine and blend, combining elements of introduced Western models of governance and elements stemming from local indigenous traditions of governance and politics, with further influences exerted by the forces of globalisation and associated societal fragmentation. They emerge from genuinely different societal spheres—spheres which do not exist in isolation from each other, but permeate each other. Consequently, these orders are shaped by the closely interwoven texture of their separate sources of origin, that is: they are hybrid (Boege et al. 2008, 2009, 2010).[17]

Under conditions of hybrid governance, the churches, as well as other locally grounded networks and associations of civil society, can act as "bridging organisation(s)" (Petzold and Ratter 2015, 40), which connect local customary life-worlds and the 'outside' world of state and international climate change policies.[18]

What is needed is the collaborative effort of such bridging organisations, non-state customary as well as civil society institutions and state institutions in planning, decision-making and implementation of climate policies. Good climate change governance depends on such collaboration. International donors, international organisations and INGOs can come into this mix in order to give financial, technical and other support.

Good climate change governance has thus to be multi-partner, multi-sector, multi-layered, integrated and holistic—not least addressing cultural and spiritual aspects which are easily underestimated by external 'Western' actors. Such integrated and holistic governance which builds on the complementarity and collaboration of all institutions and actors who are of relevance for climate change governance in hybrid political orders is essential for the prevention of climate change-induced violent conflict and for culturally sensitive as well as conflict-sensitive adaptation.

Accordingly, external actors such as international donors, international organisations and INGOs which come in with good intentions, willing to provide financial and technical support, are well advised to overcome a narrow technical and economic approach in favour of an integrated and holistic approach which pays due attention to governance issues as well as to culture and spirituality. Only then climate change policies and climate change adaptation and mitigation can be carried out in a conflict-sensitive manner.

Conflict sensitivity has to be built into any planning for climate change adaptation. Furthermore, in societal contexts where peacebuilding (post-conflict or otherwise) is of relevance, how conflict-sensitive climate change adaptation could contribute to peacebuilding has to be explored. The aim is to overcome the vicious circle of climate change and conflict through processes of mutually supportive adaptation and peacebuilding.

"Adaptation measures therefore must take into account fragility and conflict risks, while peacebuilding and conflict prevention measures need to factor in climate risks" (Taenzler, Ruettinger and Scherer 2018, 4). In the best case the outcome will be peacebuilding that supports climate change adaptation and climate change adaptation that supports peacebuilding.[19] To identify lessons learnt from peaceful and peace-supporting adaptation to the impacts of climate change and from the non-violent conduct of associated conflicts is particularly important from a policy-making and practice perspective (Adams et al. 2018).

Conclusions

What should have become clear by now is that there is no linear causal link between climate change and violent conflict. The significance of the factor 'climate change' for conflict and violent escalation of conflict has to be situated in the specific case-dependent complex conflict constellation and its emergence over time. Conflicts ensue, but whether they escalate violently is dependent on a combination of factors, "such as the intensity of the conflict over interests

and identity, the recent history of violent conflict in the receiving region etc."
(Brzoska and Froehlich 2015, 14). Governance is crucial here—whether the
effects of climate change led to violent conflict escalation or not, to a large
extent depends on the governance arrangements on the ground. What also has
become clear is that talking about governance has to include actors and institutions beyond the state. In the hybrid political orders which are predominant
in the Pacific, customary non-state actors and governance in the local community context cannot be ignored.

While the importance of non-state customary actors and institutions and
of indigenous non-Western approaches to conflict transformation and peacebuilding is increasingly acknowledged in the context of the recent 'local', 'hybrid' and 'relational turn' in peace studies (Mac Ginty and Richmond 2013,
2015; Brigg 2016; Hunt 2017), this so far has not yet filtered through to the
study of the climate change–conflict/peace nexus.

Taking these 'turns' seriously not least means to acknowledge that the people on the ground have agency of their own; they are not just passive victims of
climate change (Hermann and Kempf 2014). Local agency, however, should
not be used as an excuse for inaction of international or regional organisations
and of governments and state institutions (Pascoe 2015, 85). They have an
obligation to protect the people affected by climate change and, for example,
support those in need of relocation due to the effects of climate change.

To summarise: what is needed is a multi-layered governance framework
which encompasses actors and institutions from all levels and societal spheres.

Flowing from this insight, a few generic policy recommendations can be
drawn (see also in this volume Boege and Rakova; Higgins; Campbell; Davies).
What is needed is a holistic and integrated approach to climate change mitigation and adaptation. This comprises different dimensions:

- Acknowledging the entirety and interconnectedness of political, socio-
  economic, psychological, cultural, philosophical, spiritual, material and immaterial aspects;
- Integrating the activities of stakeholders from different societal spheres—
  state institutions, local customary as well as civil society institutions—in
  an overarching governance framework (in particular, the potential of local
  traditional actors and networks must not be left untapped—engaging them
  requires respect for their ways of operating and their worldviews);
- Taking into account the centrality of customary land tenure systems;
- Taking note of the conflict potential inherent in climate change mitigation
  and adaptation programmes and conducting such programmes in a conflict-
  sensitive way;
- Building linkages between different scales of climate change governance,
  from the international through the regional and national to the local;
- Supporting bridging institutions which have the capacity to bring together stakeholders from different governance levels, from different societal
  spheres, from different localities, and of different worldviews;

– Lastly, the industrialised countries of the Global North as the main sources of climate change have a responsibility to provide financial assistance and resources for such multi-layered climate change governance.

Notes

1 The first referendum was conducted on 4 November 2018, with 56 percent against and 43 percent for independence; however, two more referenda are to follow over the next years according to the Noumea Accord of 1998.

2 Australia and New Zealand are not understood here as being PIC, although the islands of New Zealand are geographically Pacific Islands, and Australia has a long Pacific coastline and some islands in the Pacific. Both, however, are industrialised countries of the 'first' or OECD world. This makes Australia and New Zealand clearly distinct from the other PIC. Nevertheless, both are politically (and otherwise) very active in the region and influential members of regional Pacific organisations, most importantly the Pacific Island Forum.

3 At the same time, the 22 PIC are extremely low emitters of CO_2, they only contribute a tiny 0.04 percent of the total global greenhouse gas emissions, and ranked by tonnes of CO_2 emitted per person they also are among the lowest (Foreign Affairs Committee 2010, 100).

4 Vulnerability is understood as the propensity or predisposition to be adversely affected, a system's sensitivity or susceptibility to harm and its lack of capacity to cope with the undesirable impacts of change (World Bank 2018, x).

5 Adaptation is the "process of adjustment to actual or expected climate change and its effects. In human systems, adaptation seeks to moderate or avoid harm or exploit beneficial opportunities" (World Bank 2018, vii). Adaptation, and adaptive capacity, is not only a technical issue, but has also political and social dimensions (Petzold and Ratter 2015, 36). This is an important point which will be elaborated on later in this chapter.

6 In the following, the shorthand 'climate migration' will be used for climate change-induced migration, which is "migration that can be attributed largely to the slow-onset impacts of climate change on livelihoods owing to shifts in water availability and crop productivity, or to factors such as sea level rise or storm surge" (World Bank 2018, vii).

7 For more recent stories of climate migrants see Caritas Oceania (2017).

8 As elaborated examples of such endeavours of disentangling these causal chains see Buhaug, Gleditsch and Theisen (2010), Scheffran, Link and Schilling (2012), Theisen, Gleditsch and Buhaug (2013).

9 See, for example, Theisen, Gleditsch and Buhaug (2013, 615).

10 The latter publication also reminds us, on the other hand, that "migration can be a successful and peaceful means of climate adaptation if enabled by smart policy" (Null and Rizi 2016, 5).

11 Another type of causal chain which cannot be addressed here leads from climate change adaptation and mitigation to violent conflict. On the issue of the conflict potential of climate change mitigation and adaptation see Dabelko et al. (2013).

12 Other factors contributing to the decision to migrate are the search for paid work in the cash economy, reunion of families, better education services in the capital city.

13 Gleditsch posits that "while so far there is not much evidence that robustly links climate change to major armed conflict (…), there is a more plausible argument that it may influence intergroup violence below the state level" (Gleditsch 2012, 5; see also Theisen, Gleditsch and Buhaug 2013, 622; Brzoska and Froehlich 2015).

14 "The sharpest risks emerge when the impacts of climate change overburden weak states. Climate change is the ultimate "threat multiplier": it will aggravate

already fragile situations and may contribute to social upheaval and even violent conflict" (Ruettinger et al. 2015, 1).

15 Small indigenous minorities or absolutely poor slum dwellers in general have less options to resort to large-scale violence than bigger homogenous ethnic groupings under an effective leadership and with access to weapons.

16 The 2014 IPCC assessment report in its chapter on Human Security stresses the importance of culture and cultural sensitivity, by saying that climate change threatens "cultural practices embedded in livelihoods and expressed in narratives, world views, identity, community cohesion and sense of place. Loss of land and displacement, for example on small islands and coastal communities, has well documented negative cultural and well-being impacts" (Adger/IPCC 2014, 2).

17 The concept of hybrid political order was originally developed in order to challenge the liberal peacebuilding and state-building discourse which has dominated peace and conflict studies and interventionist political practice of international organisations and leading states of the Global North in the post-Cold War era. This discourse evaluated states and peace against a reified and idealised Western notion of peace and statehood. By contrast, the hybrid political orders approach engages with the complex social, political, economic and cultural context of the intervened-upon localities and societies. It acknowledges the co-existence, competition, entanglement and blending of different types of legitimate authority and different understandings and forms of peace (formation). According to this approach, the "emergence of hybridity is the result of the clash and connection of fundamentally different forms of political organisation and community" (Richmond 2011, 19).

18 On "bridging organizations" and their role in climate change adaptation strategies, in particular with regard to connecting various actors and supporting reciprocal transfer of knowledge, see Petzold and Ratter (2015).

19 "Build pathways to peace: Identify and implement climate change programs that can support peacebuilding initiatives" (Dabelko et al. 2013, 4).

References

Adams, Courtland, Tobias Ide, Jon Barnett and Adrien Detges. 2018. "Sampling Bias in Climate-conflict Research." *Nature Climate Change Letters*. https://doi.org/10.1038/s41558-018-0068-2

Adger W. Neil, Juan M. Pulhin, Jon Barnett, Geoffrey D. Dabelko, Grete K. Hovelsrud, Marc Levy, Ürsula Oswald Spring, et al. IPCC. 2014. "Human Security". In *Climate Change 2014. Impacts, Adaptation, and Vulnerability. Part A: Global and Sectoral Aspects*. Contribution of Working Group II to the Fifth Assessment Report of the Intergovernmental Panel on Climate Change, edited by Christopher B. Field, Vincente R. Barros, David J. Dokken, Katharine J. Mach, Michael D. Mastrandrea, T. Eren Bilir, Monalisa Chatterjee, Kristie L. Ebi, Yuka O. Estrada, Robert C. Genova, Betelhem Girma, Eric S. Kissel, Andrew N. Levy, Sandy MacCracken, Patricia R. Mastrandrea, and Leslie L. White, 755–791. Cambridge: Cambridge University Press.

Boege, Volker, M. Anne Brown, Kevin Clements, and Anna Nolan. 2008. *States Emerging from Hybrid Political Orders—Pacific Experiences*. Occasional Paper No. 11. Brisbane: Australian Centre for Peace and Conflict Studies.

Boege, Volker, M. Anne Brown, Kevin Clements, and Anna Nolan. 2009. "On Hybrid Political Orders and Emerging States: What is Failing – States in the Global South or Research and Politics in the West?" *Berghof Handbook Dialogue Series No. 8*, 15–35. Berlin: Berghof Research Center for Constructive Conflict Management.

Boege, Volker, Anne Brown, Kevin Clements, and Anna Nolan. 2010. "Challenging Statebuilding as Peacebuilding - Working with Hybrid Political Orders to Build Peace." In *Palgrave Advances in Peacebuilding: Critical Developments and Approaches,* edited by Oliver P. Richmond, 99–115. New York: Palgrave Macmillan.

Boege, Volker, and Miranda Forsyth. 2018. "Customary Conflict Resolution in a State Environment: Cases from Vanuatu." In *Exploring Peace Formation: Security and Justice in Post-colonial States,* edited by Kwesi Aning, Volker Boege, M Anne Brown, and Charles T. Hunt, 191–206. London and New York: Routledge.

Brigg, Morgan. 2016. "Relational Peacebuilding. Promise Beyond Crisis." In *Peacebuilding in Crisis: Rethinking Paradigms and Practices of Transnational Cooperation,* edited by Tobias Debiel, Thomas Held, and Ulrich Schneckener, 56–69. London and New York: Routledge.

Bryant-Tokalau Jenny. 2018. *Indigenous Pacific Approaches to Climate Change: Pacific Island Countries.* Cham, Switzerland: Palgrave Macmillan.

Brzoska, Michael, and Christiane Fröhlich. 2015. "Climate Change, Migration and Violent Conflict: Vulnerabilities, Pathways and Adaptation Strategies." *Migration and Development* 5 (2): 199–210. https://doi.org/10.1080/21632324.2015.1022973

Buhaug, Halvard. 2015. "Climate-Conflict Research: Some Reflections on the Way Forward." *WIREs Climate Change* 6 (3): 269–275.

Buhaug, Halvard, Nils Petter Gleditsch, and Ole Magnus Theisen. 2010. "Implications of Climate Change for Armed Conflict." In *Social Dimensions of Climate Change: Equity and Vulnerability: New Frontiers of Social Policy,* edited by Robin Mearns and Andrew Norton, 75–101. Washington D.C.: World Bank.

Campbell, John R. 2014. "Climate-Change Migration in the Pacific." *The Contemporary Pacific* 26 (1): 1–28.

Campbell, John, and Richard Bedford. 2014. "Migration and Climate Change in Oceania." In *People on the Move in a Changing Climate,* edited by E Piguet and F Laczko, 177–204. Global Migration Issues 2. Dordrecht: Springer.

Caritas Oceania. 2017. *Turning the Tide: Caritas State of the Environment for Oceania 2017 Report.* Wellington: Caritas Aotearoa New Zealand.

Dabelko, Geoffrey D., Lauren Herzer Risi, Schuyler Null, Meaghan Parker, and Russell Sticklor, eds. 2013. *Backdraft: The Conflict Potential of Climate Change Adaptation and Mitigation.* Environmental Change and Security Program Report Vol. 14, Issue 2. Washington, D.C.: Wilson Center.

Foreign Affairs, Defence and Trade References Committee. 2010. "Security Challenges Facing Papua New Guinea and the Island States of the Southwest Pacific." Volume II. Canberra: Commonwealth of Australia, The Senate.

Gleditsch, Nils Petter. 2012. "Whither the Weather? Climate Change and Conflict." *Journal of Peace Research* 49 (1): 3–9.

Gleditsch, Nils Petter, Ragnhild Nordas, and Idean Salehyan. 2007. "Climate Change and Conflict: The Migration Link." Coping with Crisis Working Paper Series. International Peace Academy. New York: International Peace Institute.

Gleditsch, Nils Petter, and Ragnhild Nordas. 2007. "Climate Change and Conflict." *Political Geography* 26 (6): 627–638.

Hermann Elfriede, and Wolfgang Kempf. 2014. "Uncertain Futures of Belonging: Consequences of Climate Change and Sea-level Rise in Oceania." In *Belonging in Oceania: Movement, Place-Making and Multiple Identifications,* edited by Elfriede Hermann, Wolfgang Kempf, and Toon van Meijl, 189–213. New York: Berghahn Books.

Hunt, Charles T. 2018. "Relational Perspectives on Peace Formation: Symbiosis and the Provision of Security and Justice." In *Exploring Peace Formation. Security and Justice in Post-Colonial States,* edited by Kwesi Aning, M. Anne Brown, Volker Boege and Charles T. Hunt, 78–99. London and New York: Routledge.

Hunt, Charles T. 2017. "Beyond the Binaries: Towards a Relational Approach to Peacebuilding." *Global Change, Peace & Security* 29 (3): 209–227.

Ide, Tobias, Peter Michael Link, Jürgen Scheffran and Janpeter Schilling. 2016. "The Climate-Conflict Nexus: Pathways, Regional Links, and Case Studies." In *Handbook on Sustainability, Transition and Sustainable Peace* edited by Hans Günter Brauch, Úrsula Oswald Spring, John Grin, and Jürgen Scheffran, 285–304. Cham, Switzerland: Springer.

McCarter, Joe, Michael C. Gavin, Sue Baereleo, and Mark Love. 2014. "The Challenges of Maintaining Indigenous Ecological Knowledge." *Ecology and Society* 19 (3): 39–50.

Mac Ginty, Roger, and Oliver P. Richmond. 2013. "The Local Turn in Peace Building: A Critical Agenda for Peace." *Third World Quarterly* 34 (5): 763–783.

Mac Ginty, Roger, and Oliver P. Richmond. 2015. "The Fallacy of Constructing Hybrid Political Orders: A Reappraisal of the Hybrid Turn in Peacebuilding." *International Peacekeeping* 23 (2): 219–239. https://doi.org/10.1080/13533312.2015.1099440.

Nabobo-Baba, Unaisi. 2017. "In the Vanua. Personhood and Death within a Fijian Relational Ontology." In *The Relational Self. Decolonising Personhood in the Pacific*, edited by Upolu Luma Vaai and Unaisi Nabobo-Baba, 163–175. Suva: University of the South Pacific Press.

Nanau, Gordon Leua. 2017. "'Na Vanuagu': Epistemology and Personhood in Tathimboko, Guadalcanal." In *The Relational Self: Decolonising Personhood in the Pacific,* edited by Upolu Luma Vaai and Unaisi Nabobo-Baba, 177–201. Suva: University of the South Pacific Press.

Null, Schuyler, and Lauren Herzer Risi. 2016. *Navigating Complexity: Climate, Migration, and Conflict in a Changing World.* USAID Office of Conflict Management and Mitigation Discussion Paper November 2016. Washington D.C: Wilson Center.

Nurse, L., R. McLean, J. Agard, L. P. Briguglio, V. Duvat-Magnan, N. Pelesikoti, E. Tompkins, et al. IPCC. 2014. "Small Islands". In *Climate change 2014. Impacts, Adaptation, and Vulnerability. Part B: Regional Aspects.* Contribution of Working Group II Contribution to the Fifth Assessment Report of the Intergovernmental Panel on Climate Change, edited by V.R. Barros, C.B. Field, D.J. Dokken, M.D. Mastrandrea, K.J. Mach, T.E. Bilir, M. Chatterjee, K.L. Ebi, Y.O. Estrada, R.C. Genova, B. Girma, E.S. Kissel, A.N. Levy, S. MacCracken, P.R. Mastrandrea, and L.L. White, 1613–1654. Cambridge: Cambridge University Press.

Pacific Islands Forum. 2018. Forty-Ninth Pacific Islands Forum: Communique. Annex 1: Boe Declaration. Nauru, 6 September 2018.

Pascoe, Sophie. 2015. "Sailing the Waves on Our Own: Climate Change Migration, Self-determination and the Carteret Islands." *QUT Law Review* 15 (2): 72–85.

Petzold, Jan, and Beate M.W. Ratter. 2015. "Climate Change Adaptation Under a Social Capital Approach – An Analytical Framework for Small Islands." *Ocean & Coastal Management* 112: 36–43.

Reuveny, Rafael. 2008. "Ecomigration and Violent Conflict: Case Studies and Public Policy Implications." *Human Ecology* 36: 1–13.

Reuveny, Rafael. 2007. "Climate Change-induced Migration and Violent Conflict." *Political Geography* 26: 656–673.

Richmond, Oliver P. 2011. *A Post-Liberal Peace*. London and New York: Routledge.

Rüttinger, Lukas, Dan Smith, Gerald Stang, Dennis Tänzler, and Janani Vivekananda. 2015. "A New Climate for Peace: Taking Action on Climate and Fragility Risks. Executive Summary." An independent report commissioned by the G7 members. Berlin: Adelphi et al.

Salehyan, Idean. 2014. "Climate Change and Conflict: Making Sense of Disparate Findings." *Political Geography*, 43: 1–5.

Scheffran, Jürgen, P. Michael Link and Janpeter Schilling. 2012. "Theories and Models of Climate-Security Interaction: Framework and Application to a Climate Hot Spot in North Africa." In *Climate Change, Human Security and Violent Conflict*, edited by Jürgen Scheffran, Michael Brzoska, Hans Günter Brauch, Peter Michael Link, and Janpeter Schilling, 91–131. Hexagon Series on Human and Environmental Security and Peace 8. Berlin-Heidelberg: Springer.

Scheffran, Jürgen, Michael Brzoska, Jasmin Kominek, P. Michael Link, and Janpeter Schilling. 2012. "Disentangling the Climate-Conflict Nexus: Empirical and Theoretical Assessment of Vulnerabilities and Pathways." *Review of European Studies* 4 (5): 1–13.

Scheffran, Jürgen, Elina Marmer, and Papa Sow. 2012. "Migration as a Contribution to Resilience and Innovation in Climate Adaptation: Social Networks and Co-development in Northwest Africa." *Applied Geography* 33: 119–127.

Schilling, Janpeter, Sarah Louise Nash, Tobias Ide, Jürgen Scheffran, Rebecca Froese, and Pina von Prondzinski. 2017. "Resilience and Environmental Security: Towards Joint Application in Peacebuilding." *Global Change, Peace & Security* 29 (2): 107–127.

Smith, Dan, and Janani Vivekananda. 2007. *A Climate of Conflict: The Links Between Climate Change, Peace and War*. London: International Alert.

Taenzler, Dennis, Lukas Ruettinger, and Nikolas Scherer. 2018. "Building Resilience by Linking Climate Change Adaptation, Peacebuilding and Conflict Prevention." Planetary Security Initiative Policy Brief. Berlin: Adelphi.

Theisen, Ole Magnus, Nils Petter Gleditsch, and Halvard Buhaug. 2013. "Is Climate Change a Driver of Armed conflict?" *Climate Change* 117: 613–625.

Vaai, Upolu Luma. 2017. "Relational Hermeneutics. A Return to the Relationality of the Pacific Itulagi as a Lens for Understanding and Interpreting Life." In *Relational Hermeneutics. Decolonising the Mindset and the Pacific Itulagi* edited by Upolu Luma Vaai and Aisake Casimira, 17–41. Suva: University of the South Pacific Press.

Vaai, Upolu Luma and Unaisi Nabobo-Baba. 2017. "Introduction." In *The Relational Self. Decolonising Personhood in the Pacific*, edited by Upolu Luma Vaai and Unaisi Nabobo-Baba, 1–21. Suva, University of the South Pacific Press.

Vaai, Upolu Luma and Aisake Casimira. 2017. "Introduction: A Relational Renaissance." In *Relational Hermeneutics. Decolonising the Mindset and the Pacific Itulagi*, edited by Upolu Luma Vaai, and Aisake Casimira, 1–15. Suva: University of the South Pacific Press.

Vivekananda, Janani, Shiloh Fetzek, Malin Mobjörk, Amiera Sawas, and Susanne Wolfmaier. 2017. *Action on Climate and Security Risks. Review of Progress 2017*. Den Haag: Clingendael.

World Bank. 2018. *Groundswell: Preparing for Internal Climate Migration*. Washington D.C.: The World Bank.

Yamamoto, Lilian, and Miguel Esteban. 2010. "Vanishing Island States and Sovereignty." *Ocean & Coastal Management* 53: 1–9.

5   Climate Change, Migration and Land in Oceania

John R. Campbell

Introduction

Climate change is likely to affect Pacific Island people in a variety of ways. The effects of climate change in Pacific communities are likely to be most severely felt through its direct and indirect impacts on the land. However, there has been relatively little research on the impacts of climate change upon land and particularly upon people's relationships with it. The chapter is in four parts. The first part examines the meaning of land in a Pacific context to illustrate the deep connections that bond Pacific people and their land. The second part briefly outlines the likely (and in many cases already occurring) impacts of climate change on the land with a focus on how climate change is likely to impinge upon the critical relationship between people and their land. The third part then turns to population mobility as a response to these impacts including migration and an extreme form of adaptation to climate change—community relocation and displacement from customary lands. Fourth, the chapter considers the links between conflict and the effects of, and responses to, climate change and suggests possibilities for reducing the likelihood of violence as a result of climate change.

The Importance of the Land

Throughout the Pacific region people have a special relationship with their customary lands, even when they live distant from them. This relationship is encapsulated in the Polynesian term fanua:

> … in Cook Islands Maori, "enua" means "land, country, territory, afterbirth": in Futuna (Wallace) "fanua" means "country, land, the people of a place"; in Tonga, "fonua" means "island, territory, estate, the people of the estate, placenta" and "fonualoto", "grave". We can see that in some Polynesian languages, proto-fanua is both the people and the territory that nourishes them, as a placenta nourishes a baby.
>
> (Pond 1997)

DOI: 10.4324/9781003001744-7

In Fiji the cognate term is vanua. Ravuvu (1983) describes vanua as incorporating not only the land on which people live and its physical and natural resources but also the social and cultural elements of the people who are part of it. Importantly vanua underpins iTaukei (indigenous Fijian) cultural identity.

> The people of Nakorosule [a village in Fiji] cannot live without their physical embodiment in terms of their land, upon which survival of individuals and groups depends. ... Land in this sense is thus an extension of the self; and conversely the people are an extension of the land.
>
> (Ravuvu 1988, 7)

> The vanua contains the actuality of one's past and the potentiality of one's future. It is an extension of the concept of the self. To most Fijians, the idea of parting with one's vanua or land is tantamount to parting with one's life.
>
> (Ravuvu 1983, 70)

Elsewhere in Melanesia, land is held in similar regard. Bonnemaison (1984, 1) wrote regarding Vanuatu that "[t]he clan is its land, just as the clan is its ancestors". Interestingly Bonnemaison observed that the union of people and land did not proscribe mobility when he used the metaphor of the tree and the canoe to describe life on Tanna. The tree not only provided the roots which embedded the members of the kinship group in their land but also the timber from which the canoe was built to enable people to travel. However, the importance of the roots called travellers back home and much traditional (and more recent) migration was circular. Despite this mobility, land in most parts of the region could not be left fully unoccupied as this would break the critical connection with its co-constituted group.

In Solomon Islands, Andrew Te'e (2000, 2) of the Isatabu Freedom Movement (IFM) of Guadalcanal, described the land in kinship terms stretching across time as well as space:

> Land is mummy. The dogs, pigs, birds and other living creatures are our brothers and sisters. The waters of the rivers, streams and creeks are the bloods of our ancestors. The murmuring of the water are the voices of our great, great grandparents. The trees are our uncles and nephews. We use them only when they are needed to be used. The wind and its sound is the voice of Irogali repeating the story of old: that land is sacred to the souls of Isatabu. Every part of the land and all the things on it are "sacred" to me.

This statement clearly outlines the concept of land as a living relational entity with strong spiritual elements which underpin an individual's and group's identity. Banaban understanding of land is equally relational, where land, place and people (past and present) are interwoven:

> Te aba, kainga, and te rii in Gilbertese (the Kiribati language) refer, re- spectively, to the land and the people, home or hamlet, and bones. All have linguistic, human, and material forms that can be interchangeable, substituted, or used to indicate linked parts of a whole, which is the land and people together. "Te aba" thus means both the land and the people simultaneously; there is a critical ontological unity. When speaking of land, one does not say au aba, "my land," but abau, "me-land." Te aba is thus an integrated epistemological and ontological complex linking peo- ple in deep corporeal and psychic ways to each other, to their ancestors, to their history, and to their physical environment.
>
> (Teaiwa 2015, 7–8)

Clearly the term 'land' is inadequate as a descriptor of all of these things that fanua, its cognates and other Melanesian and Micronesian terms encompass. They are very difficult to translate into any one English word, and accord- ingly poorly understood in a (post)colonial context. However, using only fanua, or vanua, for example, would exclude other forms derived from proto- fanua which all differ in some ways throughout the region and the other non- cognate terms that also exist especially in Melanesia and Micronesia. Despite its shortcomings, then, the term land is used in the remainder of this chapter but its widespread and deep meaning throughout the Pacific Island region should not be forgotten.

Figure 5.1, while somewhat reductionist given the complexity of Pacific land (from time to time the clumsier term 'Pacific land' is used to remind us that it is much more than cadastral coordinates, or a collection of resources implied in the simple term 'land'), seeks to illustrate the aspects of vanua iden- tified by Ravuvu as three intersecting elements that provide physical, social and cultural security. At the centre is positioned ontological security (e.g. Giddens 1991) which enables individuals to rely on things—people, objects, places, meanings—remaining, by and large, the same tomorrow as they were today and the day before. It provides a "secure base to which [people] can return" (Hiscock et al. 2001, 50) and protection from uncertainty. It relates to a feeling of continuity in one's life that is based on a sense of belonging and confidence in one's identity (Giddens 1991). Gidden's concept of ontological security refers to individualised security in the era of late modernity. Perhaps given the complexity of Pacific land, and the importance of the kinship group (linked to land) in people's identity, the term 'more than ontological security' may be more appropriate. Figure 5.1 suggests that if any one of the three com- ponents of land is missing, then this ontological security cannot be achieved. Moreover, it also suggests that even as a result of loss of connection to land, if people can be provided with livelihoods, another site on which to build a home and work, and sustain social security (e.g. kinship networks) and even

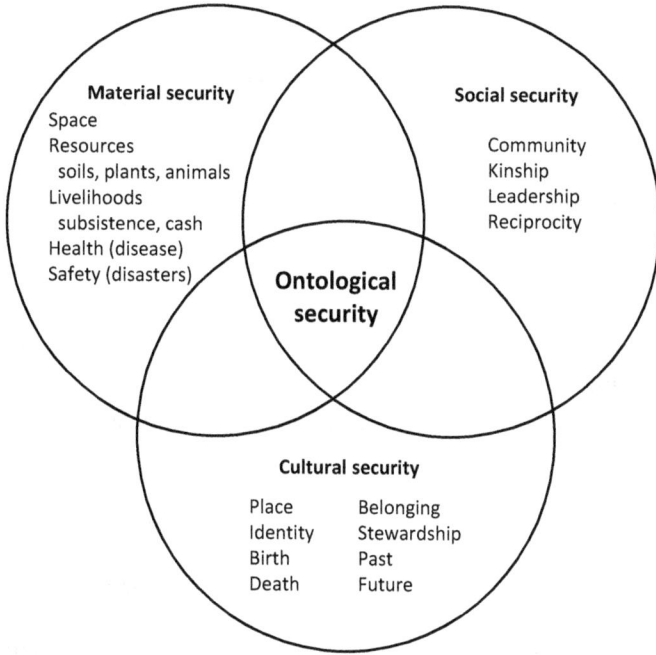

Figure 5.1 Conceptualising Pacific land as security

cultural continuity (e.g. language and other customs), ontological security will nevertheless be disrupted or lost. It is suggested that a disruption of ontological security is a form of loss and damage that is impossible to compensate.

While there can be little doubt that land is extremely important to Pacific people, it is important to acknowledge that systems of ownership and tenure vary considerably and to not fall into an "indigenous essentialism" (Hviding 1993) where land ownership is reified. Land can be, and has been through time, exchanged in a variety of traditional arrangements. People have always been mobile. Indeed, in many parts of the Pacific most people lived away from the coast (at least during the Little Ice Age from the Fourteenth to the Nineteenth Century (Nunn 2007)) and only moved to coastal settlements relatively recently as a result of missionisation and colonial control. People can also live and conduct their livelihoods on land that is not customarily theirs but usually such arrangements are based on usufruct rights (temporary rights to use the property of another) and are not permanent, although often there is confusion about such rights particularly where use of the land spans generations.

Accordingly, while we should not essentialise the mutually constitutive nature of the union between people and their land, to underestimate its importance, or worse, to completely neglect it, in relation to climate change impacts and development of responses to them would be highly problematic. The great majority of land in independent Pacific Island countries, over 90 percent

(AusAID 2008), is inalienable and cannot be bought or sold or otherwise permanently transferred (although some land is taken for public works such as airports and roads). It is generally considered to be held communally rather than by individuals, and to belong to past and future generations as well as those of the present. Rights to land are closely guarded and can be matters of tension even within (and between) otherwise cooperative communities. These issues are important when we consider how people induced or forced to resettle or relocate by climate change can find destinations that will allow durable systems of tenure. More difficult will be ways of enabling migrants to maintain their ontological security.

Climate Change and the Land

How is climate change likely to impact upon the land of the Pacific Island region? The physical impacts upon the land are relatively well understood. These include coastal erosion and inundation, increased incidence and/or frequency of extreme climatic events, changing disease vectors and other health issues, and possibly reduced agricultural and fisheries production. These impinge upon material security by damaging the spaces where people live, their resources and, in turn, their livelihoods. People are likely to face more difficult circumstances with reduced subsistence and cash incomes. Changing disease vectors, pressures on water quantity and quality, and increasing magnitude and/or frequency of extreme climatic and coastal events may reduce habitat security. What is less clear is how climate change will impact upon the non-material aspects of security.

These changes may, in turn, place pressures on social and cultural security. For example, reduced productivity may place pressure on systems of reciprocity as people are unable to fulfil their obligations. In turn, this may place pressure on the solidity of kinship systems where reciprocity plays an integral role. If migration increases, community cohesion is also likely to be compromised. While kinship systems may be relatively resilient, it is difficult to see how they would be sustained if members were spread widely, especially if a core no longer remained on the ancestral land. Nevertheless, transnational kinship networks are already emerging with nodes found not only in Pacific Island countries, but in Pacific Rim states with significant Pacific Island diasporas. In terms of cultural security, loss of physical place may severely challenge people's identity and belonging and remove critical links to their past and future. Overall it may be difficult for the traditional safety nets of Pacific communities to be sustained. Taken together, these impacts are likely to have very serious implications for the ontological security of members of Pacific Island communities.

But it is likely the biggest loss will be in relation to ontological security. Most serious will be the complete loss of customary land or significant parts of it. The embodied nature of Pacific land will be destroyed and if this happens, as Ravuvu (above) states, it would be equivalent to the loss of life. Kinship groups whose identity is embedded in the land would lose an important rationale for their existence and it is likely that there will be significant emotional and spiritual losses. Even if the land remains, but is uninhabitable, the continuity of the relationship will be broken, again with very serious psychological consequences. If climate change response strategies and projects continue to be dominated by international agencies informed by 'western' scientific, engineering and economic practices, it is likely that these important losses will be neglected.

Migration and Land

The links between climate change and migration are covered elsewhere in this policy brief series. However, it is important to briefly consider aspects of climate change mobility as they relate to the land issue. Two broad categories of climate change mobility can be identified as shown in Table 5.1 (Campbell 2014). First, climate change induced migration is likely to be linked to environmental degradation at the place of origin such that not all inhabitants can be satisfactorily supported by the local resource base. Some are induced to migrate, thereby relieving population pressure on the dwindling local resources and at the same time augmenting the livelihoods of those who remain through remittances. Migrants such as these may return home (permanently or temporarily), if they wish, and maintain their links to their land (even after several generations away from it), providing the connection with the land is sustained by the people who remain. Such induced migration may already be occurring although it is difficult to distinguish it from other reasons for migration in the region. For many communities, satisfying contemporary needs and wants is difficult if they depend only on local land-based resources, hence the current high rates of urban and external migration. Climate change is likely to exacerbate this process. Climate change induced migrants who remain within their countries will need to find a place to live and conduct their livelihoods. Some may seek rural locations (perhaps through kinship including affinal connections) where they may obtain usufruct rights to use land. However, if the number of migrants increases, or they increase their demands on the land and/or seek to expand their usage, tensions can develop (Allen 2012). It is likely, though, that most induced migrants are likely to find themselves in urban or peri-urban squatter settlements within their own countries where similar problems relating to land may emerge (much urban, and particularly peri-urban, land in the region is held under customary forms of ownership).

Several Pacific Island countries (PICs) have migration access to Pacific Rim countries, based mostly on current or previous colonial status. American territories such as Guam, CNMI and American Samoa have access to the USA as do Palau, FSM and the Marshall Islands under their compacts of free association with the USA. Niue and Cook Islands have similar relations with New Zealand, and Samoa has a special relationship as a former colony. Fiji and Tonga also have significant migration flows into New Zealand. French Territories have access to France. The former UK colonies of Solomon Islands,

Table 5.1 Climate change migration and relocation options

Destination		CC Induced Migration	CC Forced Relocation
Internal	Proximate	Within customary lands	Individual and family migration, much along existing patterns (internal urban migration and external migration). Few problems.	Large proportion of community resettled but sustaining similar sets of social and cultural activities as in place of origin.
		Neighbouring lands (non-customary)	Land issues need to be resolved, possible usufruct rights, possible affinal kinship linkages for example	The least problematic option and historically relatively common. Land issues need to be resolved; success may depend on relationships with neighbouring communities
	Distant	Rural area elsewhere in country	Land issues need to be resolved, possible usufruct rights, possible affinal kinship linkages for example	Land issues need to be resolved. May be very difficult in case of distant relocations
		Urban Areas	Informal settlements, usufruct rights, unemployment and underemployment	Informal settlements, usufruct rights, many settlements in PIC urban areas comprise people from similar places of origin.
External	Other PIC	Other PIC	Most likely urban, as above	Land issues need to be resolved. Possible cultural and language difficulties and adjustment to very different environments
	Beyond PICs	Most likely Pacific Rim countries (NZ, Aust, USA?)	Most likely urban rental or temporary agricultural labour schemes	Difficult to envisage recreation of community social and other structures in non-PIC settings

Vanuatu (as Condominium of the New Hebrides with France), Tuvalu and Kiribati have much more limited opportunities for external migration as does Papua New Guinea (formerly an Australian colony). Indeed, it is the three Melanesian countries (Papua New Guinea, Solomon Islands and Vanuatu) that are facing some of the most serious issues relating to internal migration and urbanisation, while the two atoll countries have some of the most serious scenarios for loss of land, livelihood and habitat security. Kiribati in particular is heavily impacted by urban migration, with South Tarawa exhibiting among the highest population densities of any part of the Pacific region. Drought induced urban migration is also a concern in Papua New Guinea, where the densely populated Highlands region is subjected to periodic droughts associated with El Niño events that have occasioned massive relief efforts, partly to reduce the numbers heading to urban destinations (Campbell and Warrick 2014). There is insufficient certainty, but it is possible the El Niño Southern Oscillation phenomenon may be affected by climate change (Australian Bureau of Meteorology and Commonwealth Scientific and Industrial Research Organisation 2014).

The second form of mobility, climate change forced relocation, will result in cases where the land is no longer able to support its people. This may be through the loss of material, social and cultural security (as shown in Figure 5.1). In these cases, there is likely to be little choice but for the whole community (or at the least, a very large portion of it) to find a place where these elements of security can be sustained. This is likely to be the most difficult of the mobility outcomes of climate change as it involves the loss of the land by the relocating community. While aspects of security can be recovered, the bases of ontological security are likely to be much more difficult to re-establish in the short term (which is considered to be several generations). In addition, relocation will involve other communities giving up parts of their own land, which will also impact upon their own sense of cultural identity and security. Where sufficient land cannot be found to support not just housing but also livelihoods, it is likely that communities will break up and individual families or groups will make independent migration decisions, with many finding themselves in urban locations, with no land to which to belong, or return.

Resettled Migrants

While much conflict in PICs, especially Melanesia, has been described as ethnic conflict, land has also often been seen as one, if not the, underlying cause of tensions leading to violence. Of course, the reasons for conflict are often complex, rooted in uneven development, internal migration, unequal levels of access to resources and the like. Migrants are often restricted in their access to land; for those 'settlers' who have access it is often unstable; and land has emerged as the basis for a politics of exclusion based on indigeneity (Allen 2012; Droogan and Waldek 2015).

Two case studies from Melanesia help inform us of issues that may arise when there is significant in-migration of 'outsiders'. In Papua New Guinea (PNG) and Solomon Islands, tensions have emerged between local people and migrants from elsewhere in the respective countries which have seen divisions emerge between landowners and settlers.

In the PNG case, migration was to oil palm land settlement schemes on land alienated from local owners by the PNG government (Koczberski and Curry 2004). Tensions emerged and landowners demanded compensation; at times violence erupted between increasingly polarised landowners and settlers. However, Curry and Koczberski (2009) have reported that systems of tenure have emerged that enable 'outsiders' to continue to use land for long-term crops through indigenous (gift) exchange which sustains social relations between the parties involved, based on customary principles of land tenure in which the land is inalienable. The arrangements, however, do require reciprocal obligations to be met by the occupiers.

> The indigenous morality of gift exchange means that the more frequent and intense previous gift giving has been, the stronger is the moral basis of an outsiders land claim and the more difficult it is for the customary landowners to reclaim it.
>
> (Curry and Koczberski 2009, 104).

Such measures are not necessarily permanent though. When parties to the arrangements die (either the owning clan leader or outsider) or oil palms need to be replanted (20–25 years), the social relations involved may be re-evaluated. This generational uncertainty may be resolved by establishing new sets of social relationships and so on, although as conditions change through time this may become difficult (Curry and Koczberski 2009). In the long-term, durable relationships may be difficult to achieve.

The Solomon Islands case relates to tensions that have arisen as large numbers have migrated in recent decades, particularly from Malaita to Guadalcanal (Allen 2012; Droogan and Waldek 2015). Here the conflict that emerged resulted in hundreds of deaths and thousands of people being displaced (Allen 2006), requiring the intervention of a Pacific regional force to quell the disruption. One of the reasons for the migration from Malaita to Guadalcanal was the differences in development opportunities on the two islands. Initially relatively small numbers of migrants were able to obtain access to land from the customary owners, but tensions resulted as the numbers of settlers increased (Allen 2012). Part of the problem was that the settlers and landowners viewed land tenure differently. Moreover, Allen (2012, 169) observed "salient generational factors" that led to breakdowns not only between landowners and settlers but also among the landowners themselves, such that original arrangements broke down. An underlying concern was the failure of subsequent settlers to pay sufficient respect to the customary owners.

In both cases the tension and conflict that occurred has been categorised as inter-ethnic violence, but the reasons are considerably more complex and include issues related to land ownership. Moreover, they indicate that tensions appear to emerge only when the number of settlers increases, and as generations pass, levels of respect between the parties seem to decline.

How do these cases inform us about climate change migration? First, they indicate that land is an important underlying issue, but it is important to note that it is one of several causative factors. Second, underpinning the migration in these countries was uneven development, with people from disadvantaged areas seeking better economic opportunities elsewhere. If the effects of climate change were to curtail development opportunities, similar migration patterns may arise elsewhere in the Pacific. Moreover, the case studies tend to suggest that the potential for conflict increases as the number of settlers grows, either in rural or urban areas. It is likely that climate change may well lead to increasing numbers of migrants throughout the region, and a possible increase in tensions between settlers and land owners. While there may be initial sympathy extended to people migrating from environments degraded by climate change, this may change through time. Mechanisms will need to be developed that ensure relations between landowning groups and settlers remain respectful.

Relocated Communities

While climate change-induced internal migration may lead to increased levels of conflict in the region, the most difficult situations are likely to arise when whole communities are forced to leave their lands. First, as noted, this would break the continuity of the bond between people and their land, and in doing so, destroy their ontological security; the material, social and cultural aspects of security would also be compromised. Second, to overcome these losses, large tracts of land would need to be made available to enable the communities to have space on which to live. This would also require the connection between another group and part of its land to be broken.

There have been numerous relocations of villages in PICs over the ages, especially to locations within customary lands and, to a lesser degree, to nearby lands belonging to neighbouring kinship groups, often involving customary forms of exchange. Many such moves have taken place after devastating disasters such as tropical cyclones including storm surges or flash flooding (Campbell, Goldsmith and Koshy 2007). Much less common has been the relocation of communities to distant locations either within countries or internationally. It is possible, particularly in the case of atoll communities, that long-distance relocation in very different environments may be required.

An informative case is that of the relocation of the people of Banaba (a raised atoll devastated by phosphate mining) located in what is now Kiribati, to Rabi, an island in north-eastern Fiji (Teaiwa (2015) provides an insightful and detailed account on the relocation of the Banaban people to Rabi). The relocation was encouraged (if not imposed) by the colonial government of the Gilbert and Ellice islands ostensibly to resettle Banabans affected by Japanese occupation in the Second World War, and to transfer many Banabans to occupied islands elsewhere in Micronesia as labour; probably more likely, there was a desire by the British Phosphate commission to have unfettered access to further exploit the phosphate on the island. The relationship that the Banaban people have with their land is described as blood being mixed with the land and conversely the land is in the people's blood (Teaiwa 1998; Silverman 1971). Rabi is an island that is considerably larger in area (ten times the size) and more fertile than Banaba (especially after it was denuded by the phosphate mining). It is perhaps not surprising that after two years the transplanted people on Rabi voted to stay on their new island. But despite this, there were misgivings over their separation from Banaba, heightened by injustices in the distribution of royalties from the phosphate, most of which went initially to the colony and later to the newly independent country of Kiribati. The Banabans want Banaba to be independent of Kiribati and have placed a caretaker population on Banaba to ensure their continued connection to the island. The situation for the Banabans remains ambiguous after almost three-quarters of a century of relocation. While on the one hand seeking to retain their bond with Banaba, their place in Fiji is also a source of concern. Fraenkel (2003) reported that the Banabans are among the most marginalised communities in Fiji. In addition, the original inhabitants of Banaba are seeking to redress their separation from their vanua (Campbell 2010). As Teaiwa (2015, 19) observes, Banaba is paradoxically a contested homeland:

> Many of the descendants of the iTaukei inhabitants of Rabi … live on islands surrounding Rabi and have maintained strong ancestral links to their home island. This has made for some very awkward interactions between Banabans and Fijians, and Rabi is thus a still contested place with two displaced populations who call it home.

Adding further irony, some atoll-based i-Kiribati, concerned about how they might respond to climate change and sea-level rise, see Banaba (highest elevation 81m.) as a possible site for relocation should their low-lying islands become uninhabitable (Corcoran 2016).

Other examples are the relocation of Carteret Islanders and the resettlement of 'Gilbertese' people. In the former case there have been attempts to find land on nearby Bougainville Island for the people of the Carteret islands which are being affected by both subsidence and rising sea-levels (Boege and Rakova 2019). This is an example of internal relocation but beyond the relocating group's customary lands. Despite these efforts over several decades, there has been very limited success. While land for housing has been found, local land owners have been less able to make more land available for relocated people to carry out their livelihood activities. In the second case 'Gilbertese people' were resettled, under the same colonial regime that moved the Banabans, after perceptions that there was overpopulation on their atolls. Initially relocated in the Phoenix islands, the resettlement failed largely due to the arid conditions on these formerly uninhabited atolls. The group was then transferred to the Western Province of Solomon Islands but had limited access to resources and land. There was resentment from local leaders that the Gilbertese resettlement was in Western Province and not spread elsewhere in Solomon Islands (Fraenkel 2003; Premdas, Steeves and Larmour 1984). With limited access to land, Gilbertese settlements were located close to the ocean and Gilbertese people were over-represented in the fatalities from the 2007 tsunami and faced difficulties in recovering from the event (Donner 2015).

Even relocation to neighbouring lands can be problematic. Cagilaba (2005) reports on a coastal community in Fiji that was relocated upslope after being affected by tropical cyclones in the 1970s. The new site was partly on land belonging to a neighbouring mataqali (clan) but the move was made possible by a traditional arrangement between the two communities. Three decades later, younger members of the mataqali that ceded the land claimed the land back as the exchange was not recorded in the official land records. This was only a portion of the relocated land but was enough for grievances to arise. While traditional systems of land tenure are often described as flexible, codification and mapping of land ownership which started during the colonial era has fixed boundaries that in the past may have been considerably more elastic.

Climate Change Mobility and Land: Implications for Peace

Climate change-induced migration and forced relocation have considerable potential to increase the incidence of violent events in the Pacific Island region given the strong linkages between conflict and land, and the implications of climate change-generated mobility and displacement for people's relationships with land. But violence and conflict need not be inevitable. As we have seen, there were traditional ways in which land issues could be resolved, and with advanced notice of the likely effects of climate change, we can also plan for a future when mobility may be a significant response.

A way Forward

In an ideal world, forced community relocation would not be necessary but international negotiations to reduce greenhouse gas emissions have been painfully slow and inadequate. The impacts of climate change are likely to be particularly harsh on small islands (Nurse et al. 2014; IPCC AR5) and continued occupancy of some places may become unviable. Despite this, it needs to be stressed that community relocation, particularly beyond customary lands, as an adaptive response to climate change should be a last option and is likely to be fraught with difficulties. But, if there is no choice, such relocation will have to go ahead. Given the difficulties, such forced movement of entire communities will need to be very carefully planned. Thus paradoxically, while it should be a last option, planning for relocation should not be left to the last minute which would be likely to result in negative outcomes and could be referred to as maladaptation. Accordingly, a proactive, long-term planning approach is advocated. Reactive and inadequately planned relocations may result in problems that may fester for generations. The following is a very tentative outline of measures that may reduce the negative impacts of community relocation for both those who are forced to move and those on whose customary land they settle.

General Principles

The following list is drawn from a Submission to the Executive Committee of the Warsaw International Mechanism for Loss and Damage associated with Climate Change Impacts of the United Nations Framework Convention on Climate Change (Campbell 2017).

1 Land at risk will initially need to be identified. Many PICs have already conducted vulnerability assessments and there is growing awareness of places that may become uninhabitable.
2 People should be assisted, through support for adaptation, to stay on their customary lands as long as is possible. Costs of adaptation should not be used to avoid measures that otherwise would be effective.
3 The relocated community, once established, should be able to enjoy, in as far as it is possible, livelihood, land and habitat security as was the case in their customary homelands. This would include enough land (and fisheries) for establishing a settlement and meeting subsistence and commercial requirements.
4 Breaking the connection between a community and its land can be avoided if 'caretakers' could be supported to stay on otherwise uninhabitable land. This could be facilitated by provision of supplies such as food and water if livelihoods are fully compromised by climate change and the maintenance of existing, or establishment of new, transport connections.
5 Anticipatory planning incorporating a number of steps to keep to a minimum cultural, social, emotional, psychological disruption, and hence the potential for conflict.

   a Not only land at risk needs to be identified (see 1.), but also relocation sites. This is likely to be a fraught process as most communities will be unable to easily cede their land to others. In some settings, communities that have existing relationships with vulnerable communities may be able to begin discussions about making land available. In others, it may be possible to call for voluntary offers of land. Such an approach will be difficult but not impossible and already the Prime Minister of Fiji and the Pacific Council of Churches have acknowledged the sense of unity among the people of the region in supporting a Pacific regional 'resolution' to the issue of forced relocation.

b Early planning and consultation between people from both origin and possible destination communities. This will need to be conducted in a sensitive and culturally appropriate manner. It is important that parties understand that the planning is long-term and contingent on climate change effects worsening.

c Discussions among governments, possible relocatees and people of the land, part of which may be used for relocation, covering such issues as compensation (market, non-market and traditional forms) and the requirements of the resettled communities.

d Early interaction between representatives of the origin and destination communities including reciprocal visits to each other's land in order for both groups to understand the social, cultural and physical characteristics of each community.

e Initial resettlement by an 'advance' party of relocatees to facilitate relocation when and if required.

f Establishment of the site including building of homes, preparing gardens, establishing infrastructure. These activities should be conducted in ways that are appropriate to both the 'host' community and the relocatees (in as much as is possible).

g Resettlement.

h Monitoring and evaluation involving communities from both origin and destination. Any problems should be identified and resolved as quickly as possible.

Conclusions

This chapter has sought to identify the strong linkages between people, land, climate change, migration and possibly conflict. Central to this interconnected system is Pacific land which to date has rarely been the subject of analysis in relation to the impacts of climate change and the development of appropriate adaptive (and indeed mitigation) responses. At the heart of this is the ontological security of individuals and groups that are strongly rooted in the land. Two major issues arise. First all attempts must be made to achieve in situ adaptation so that people's essential links to their land can be sustained. Second, where relocation is unavoidable, all efforts must be made to reduce the disruption and psychological, spiritual and emotional losses that are likely to unfold. Unless further attention is paid to this issue, it is distinctly possible that some responses to climate change may cause tensions and contribute to violent outcomes.

References

Allen, Mathew. 2006. "Contemporary Histories of the Conflict in Solomon Islands." Review Works: Happy Isles in Crisis. The Historical Causes for a Failing State in Solomon Islands by Clive Moore and the Manipulation of Custom. From Uprising to Intervention in the Solomon Islands by John Fraenkel. *Oceania* 76 (3): 310–315.

Allen, Mathew. 2012. "Land, Identity and Conflict on Guadalcanal, Solomon Islands." *Australian Geographer* 43 (2): 163–180.

AusAID. 2008. Making Land Work: Volume 1: Reconciling Customary Land and Development in the Pacific. Canberra: Commonwealth of Australia.

Australian Bureau of Meteorology and Commonwealth Scientific and Industrial Research Organisation. 2014. *Climate Variability, Extremes and Change in the Western Tropical Pacific: New Science and Updated Country Reports. Pacific-Australia Climate Change Science and Adaptation Planning Program Technical Report.* Melbourne, Australian Bureau of Meteorology and Commonwealth Scientific and Industrial Research Organisation.

Boege, Volker, and Ursula Rakova. 2019. "Climate Change-Induced Relocation: Problems and Achievements - The Carterets Case." Toda Peace Institute Policy Brief 33. Tokyo: Toda Peace Institute. Available at https://toda.org/policy-briefs-and-resources/policy-briefs/climate-   change-induced-relocation-problems-and-achievements-the-carterets-case.html

Bonnemaison, Joel. 1984. "Social and Cultural Aspects of Land Tenure." In *Land Tenure in Vanuatu*, edited by Peter Larmour, 1–5. Suva: University of the South Pacific.

Cagilaba, Vinau. 2005. "Fight or Flight?: Resilience and Vulnerability in Rural Fiji." MSocSc thesis, University of Waikato.

Campbell, John R. 2010. "Climate-Induced Community Relocation in the Pacific: The Meaning and Importance of Land." In *Climate Change and Displacement: Multidisciplinary Perspectives*, edited by Jane McAdam, 57–79. Oxford: Hart Publishing.

Campbell, John R. 2014. "Climate-Change Migration in the Pacific." *The Contemporary Pacific* 26 (1): 1–28.

Campbell, John R. 2017. "The Implications of Climate Change for the Loss and Damage Caused by Disruption of the Essential Link between People and Their Land." Submission to the Executive Committee of the Warsaw International Mechanism for Loss and Damage associated with Climate Change Impacts of the United Nations Framework Convention on Climate Change. https://unfccc.int/files/adaptation/groups_committees/loss_and_damage_executive_committee/application/pdf/l_d_submission_j__campbell.pdf

Campbell, John R., Michael Goldsmith, and Kanyathu Koshy. 2007. "Community Relocation as an Option for Adaptation to the Effects of Climate Change and Climate Variability in Pacific Island Countries (PICs)." Final report for APN project 2005-14-NSY-Campbell. Tokyo: Asia Pacific Network for Global Change Research.

Campbell, John R., and Olivia Warrick. 2014. *Climate Change and Migration Issues in the Pacific*. Suva: United Nations Economic and Social Commission for Asia and the Pacific (ESDCAP).

Corcoran, John. 2016. "Implications of Climate Change for the Livelihoods of Urban Dwellers in Kiribati." PhD Thesis, University of Waikato.

Curry, George N., and Gina Koczberski. 2009. "Finding Common Ground: Relational Concepts of Land Tenure and Economy in the Oil Palm Frontier of Papua New Guinea." *The Geographical Journal* 175 (2): 98–111.

Donner, Simon D. 2015. "The Legacy of Migration in Response to Climate Stress: Learning From the Gilbertese Resettlement in the Solomon Islands" *Natural Resources Forum* 39 (3–4): 191–201.

Droogan, Julian, and Lise Waldek. 2015. "Continuing Drivers of Violence in Honiara: Making Friends and Influencing People." *Australian Journal of International Affairs* 69 (3): 285–304. DOI: 10.1080/10357718.2014.992859.

Fraenkel, Jon. 2003. "Minority Rights in Fiji and the Solomon Islands: Reinforcing Constitutional Protections, Establishing Land Rights and Overcoming Poverty." Paper prepared for United Nations Commission on Human Rights, Sub-Commission on Promotion and Protection of Human Rights, Working Group on Minorities. UN Doc E/CN.4/Sub.2/AC.5/2003/w.p.5.

Giddens, Anthony. 1991. *Modernity and Self-identity: Self and Society in the Late Modern Age.* Cambridge: Polity Press.

Hiscock, Rosemary, Ade Kearns, Sally MacIntyre, and Anne Ellaway. 2001. "Ontological Security and Psycho-Social Benefits from the Home: Qualitative Evidence on Issues of Tenure." *Housing, Theory and Society* 18 (1): 50–66.

Hviding, Edvard. 1993. "Indigenous Essentialism? 'Simplifying' Customary Land Ownership in New Georgia, Solomon Islands." *Bijdragen tot de Taal-, Land- en Volkenkunde, Politics, Tradition and Change in the Pacific* 149 (4): 802–824.

Koczberski, Gina, and George N. Curry. 2004. "Divided Communities and Contested Landscapes: Mobility, Development and Shifting Identities in Migrant Destination Sites in Papua New Guinea." *Asia Pacific Viewpoint* 45 (3): 357–371.

Nunn, Patrick. 2007. *Climate, Environment and Society in the Pacific during the Last Millennium.* Amsterdam: Elsevier.

Nurse, L., R. McLean, J. Agard, L. P. Briguglio, V. Duvat-Magnan, N. Pelesikoti, E. Tompkins, et al. IPCC 2014. "Small Islands". In *Climate Change 2014. Impacts, Adaptation, and Vulnerability. Part B: Regional Aspects.* Contribution of Working Group II Contribution to the Fifth Assessment Report of the Intergovernmental Panel on Climate Change, edited by V.R. Barros, C.B. Field, D.J. Dokken, M.D. Mastrandrea, K.J. Mach, T.E. Bilir, M. Chatterjee, K.L. Ebi, Y.O. Estrada, R.C. Genova, B. Girma, E.S. Kissel, A.N. Levy, S. MacCracken, P.R. Mastrandrea, and L.L. White, 1613–1654. Cambridge: Cambridge University Press.

Pond, Wendy. 1997. "The Land as a Traceable Commodity." *New Zealand Books,* 7 (5) Issue 31, Spring. Posted on December 12, 1997. https://nzbooks.org.nz/1997/contents/issue-31-contents/

Premdas, Ralph R., Jeffrey S. Steeves, and Peter Larmour. 1984. "The Western Breakaway Movement in the Solomon Islands." *Pacific Studies* 7 (2): 34–67.

Ravuvu, Asesela. 1983. *Vaka i taukei: The Fijian Way of Life.* Suva: University of the South Pacific.

Ravuvu, Asesela. 1988. *Development or Dependence: The Pattern of Change in a Fijian Village.* Suva: University of the South Pacific.

Silverman, Martin G. 1971. *Disconcerting Issue. Meaning and Struggle in a Resettled Pacific Community.* Chicago: University of Chicago Press.

Teaiwa, Katerina. 2015. *Consuming Ocean Island: Stories of People and Phosphate from Banaba.* Indiana: Indiana University Press.

Teaiwa, Teresia. 1998. "Yaqona/Yagoqu: The Roots and Routes of a Displaced Native." *UTS Review* 4 (1): 92–106.

Te'e, Andrew. 2000. "Land is Sacred to Me (Part One)." *Isatabu Tavuli* 1 (4): 2 http://lukluksi.tripod.com/TAVULI4.htm. Accessed 12/02/2019.

Part III

Case Studies

Climate Change and Conflict
in the Pacific

6 Relocation of Carteret Islanders to Bougainville

A Special Case of Climate Change Adaptation

Volker Boege and Ursula Rakova

Introduction

The effects of climate change force governments and people in Pacific Island countries (and elsewhere) to think about, plan for and implement a variety of adaptation measures. Planned relocation of particularly affected island communities is increasingly being discussed as such an adaptation measure in cases where islanders are confronted with severe challenges to land security, livelihood security and habitat security (Campbell 2014, 4–5).

People try to adapt to changing conditions. But often options for *in situ* technical adaptation—planting mangroves, building seawalls, setting up rainwater tanks—are limited. Often, they are technically not feasible or too costly, or they only work as interim measures. Consequently, migration or resettlement might be the better, or even the only, option of long-term sustainable adaptation (Boege 2018).

Community relocation "may be considered the most extreme form of climate migration and is considered by many to be a last-resort adaptation option" (Campbell 2014, 11). It is different from previous resettlement of communities (which has occurred in Oceania on many occasions in the past); under conditions of climate change it will become much more widespread and, even more importantly, there will be no return option. You cannot return to a sunken island or an island that has become uninhabitable. But there will be time for relatively long-term planning, at least with regard to the slow-onset effects of climate change. Such planned relocation is caused by the insight that there are no other options left, at least not long-term, and thus there is an element of 'forced' to it. On the other hand, planned relocation is largely 'voluntary'; people or their political leaders take decisions regarding relocation under terms and conditions that they can influence themselves, at least to a certain extent. They are not just victims of forced displacement (like people who have to move due to natural disasters or major economic projects, e.g. dams or mines).

Today there is a lot of talk in Pacific Island countries about the need to relocate, often rather alarmist and sensationalist. But there is much less planning for relocation, and even less actual relocation happening. Arguably, one of the

DOI: 10.4324/9781003001744-9

most advanced (and widely publicised) climate-related relocation programmes in the Pacific to date is the resettlement of Carteret Islanders from their atoll to mainland Bougainville in Papua New Guinea (PNG). In this chapter we will take a closer look at this case, because it encompasses a broad range of issues which will become relevant for other relocation endeavours elsewhere and in the future. It can be seen as a paradigmatic case.

The chapter is structured as follows: first, a brief overview of the current situation on the Carteret Islands and on Bougainville is given. This is followed, second, by the presentation of the Carterets Integrated Relocation Program and its implementation. Third, the highly important land issue is given particular attention. The fourth section addresses the (lack of) external support. This is, fifthly, followed by a discussion of a range of problems associated with previous and current state-led relocation attempts, and the associated concerns and expectations of affected people. Finally, conclusions are drawn, and, flowing from that, a couple of policy recommendations are made.

The Carteret Islands

The Carterets atoll comprises six low-lying islands (Han, Huene, Iangain, Yesila, Yolasa, Piul), with a combined land area of just 0.6 square kilometres, inhabited by approximately 3,000 people. This makes the Carterets the island group with the highest population density in PNG (Edwards 2013, 71). The islands belong to the Autonomous Region of Bougainville (ARoB), which is part of PNG.[1] The Carterets are located 86 kilometres northeast of the main island of Bougainville, a four-hour ride by banana boat.

With a maximum elevation of 1.2 metres above sea level and due to their high ratio of coastline to surface land area, the islands are particularly vulnerable to sea level rise, high tides and storm surges. Coastal erosion affects shorelines. King tides have become a regular occurrence and are becoming more and more dangerous. In February 2018 a king tide swept over Yesila Isle, right through family homes and food gardens. The people have great difficulties maintaining their subsistence economy, which is based on fish, bananas, taro and other vegetables, grown in food gardens. Swamp taro, which used to be the main staple food crop, cannot grow any more due to salt water intrusion and salinisation of soil and water. Today, only banana and coconut can grow on the islands (Edwards 2013, 73). Hence the "Carteret Islanders story is an 'early warning' to all of the threat posed to food supplies by climate change" (Caritas 2018, 39). Soils become more and more swampy, providing breeding grounds for mosquitos; as a consequence, malaria becomes more frequent (Edwards 2013, 74). Freshwater wells have been contaminated by saltwater, making freshwater more and more scarce. Food and water security are threatened (Rakova 2014; Tulele Peisa, n.d.). The export of farmed seaweed to Asian markets and the capture of bech-de-mer (sea cucumber) provide some cash income for the islanders, and some families receive remittances from relatives working on Bougainville or elsewhere in PNG (Edwards 2013, 68; UNDP 2016, 5).[2]

People are becoming increasingly dependent on food aid shipped in from mainland Bougainville (Edwards 2013, 73); the diet, however, is unhealthy (rice and flour) and the shipments are irregular and unreliable. Travelling between the islands and mainland Bougainville is getting more dangerous as there are more and more severe storms happening almost all year round (Caritas 2018, 25). Other atolls in the ARoB are in a similar situation, namely Tasman, Mortlock and Nuguria, which have a total population of about 2,500.

In situ adaptation is difficult and only modestly successful: raised bed gardens (supsup gardens), mangrove seed planting, water tanks, and seawalls. Seawalls on Han Island are broken. Mangrove roots cannot grip and hold as the sea currents are very strong and continuously remove the mangrove sediments to protect the shorelines (Connell 2018).

Given these conditions, relocation to the main island of Bougainville is another adaptation option for the Carteret Islanders.

Destination Bougainville

Bougainville is currently in a difficult stage of its history. For almost ten years (1989–1998) it was the theatre of a war of secession between the Bougainville Revolutionary Army and the security forces of the national government of PNG (and Bougainville auxiliaries). The war was the longest and bloodiest violent conflict in Oceania since the end of the Second World War. Over the last two decades Bougainville has undergone a comprehensive process of post-conflict peacebuilding, which has been relatively successful. Currently Bougainville is an autonomous region within PNG, with its own constitution and its own government, the Autonomous Bougainville Government (ABG). The establishment of the ARoB and the ABG was one of the two main political pillars of the Bougainville Peace Agreement of August 2001. The other one was the guarantee of a referendum on the future political status of Bougainville: either complete independence or remaining with PNG. The BPA stipulates that the referendum has to be held ten to fifteen years after the establishment of an autonomous government for Bougainville (which took place in 2005). Hence the window for holding the referendum was from June 2015 to 2020.

In May 2016 the ABG and the Government of PNG agreed upon 15 June 2019 as the target date for the referendum. From then on, the focus of politics and public attention in Bougainville was clearly on the upcoming referendum. Due to delays in preparations, the target date of 15 June could not be met. But the referendum was conducted from 23 November 2019 to 7 December 2019. On 11 December the result was officially declared. Bougainvilleans overwhelmingly opted for independence (97.7 percent). This, however, does not mean automatic independence. The BPA stipulates that the referendum is non-binding, that the two governments have to consult over its outcome, and the PNG national Parliament has to ratify it. Hence there is a need for a transition process after the referendum. This process is expected to last for a couple of years. It can be argued that only with the peaceful implementation of the referendum result, will Bougainville peacebuilding have reached a satisfactory conclusion. Bougainvilleans today face the challenge of building a polity which is effective and legitimate in maintaining sustainable peace and order. Currently, the focus of the ABG and the general public in the ARoB is intensely on the transition. For the time being, all other issues, including the effects of climate change and the need for relocation of climate change affected communities, have been put on the back burner. The government and the people have to work hard to secure a peaceful post-referendum transition phase. Such transition by its nature is fragile and burdened with a broad spectrum of challenges. Moreover, the situation is still volatile in parts of the island; there are issues with regard to localised violence, and governance problems like corruption etc. Land is scarce on Bougainville, and conflicts over land are common. Hence mainland Bougainville today has its challenges as a destination of resettlement for Carteret Islanders.

Relocation—Tulele Peisa and the Carterets Integrated Relocation Program

Despite the fragile situation on Bougainville, Carteret Islanders intend to resettle there. Due to the effects of climate change they do not see a future for themselves and their children on their home islands. People from the Carterets themselves took the initiative to develop a relocation plan. After a series of community meetings which discussed the worsening situation on the atolls, the Carterets Council of Elders (CoE), the local governing body on the islands, in late 2006 decided to form an NGO to organise the resettlement. The organisation was named 'Tulele Peisa', which in the local halia language means 'sailing the waves on our own'. "This name choice reflects the elders' desire to see Carteret islanders remain strong and self-reliant" (Rakova 2009, 2). Tulele Peisa elaborated a detailed 20-step relocation plan, the Carterets Integrated Relocation Program (CIRP) with the aim of voluntary relocation of approximately 1,700 Carteret Islanders to several locations on mainland Bougainville: Tinputz, Tearouki and Keriaka gifted by the Catholic Diocese on humanitarian grounds, and Wakunai and Tenapo which are family plantations privately owned by two Carterets families. Negotiations are underway to legally acquire Wakunai and Tenapo from the families to relocate more Carterets families.

Resettlement land at the first three sites was provided by the Catholic Church (talks with landholding groups in North and Central Bougainville about the provision of land for Carterets relocatees had led nowhere). This came after the Bishop of the Catholic Diocese of Bougainville had visited the Carterets for two weeks in January 2007 and had "witnessed for himself the dire straits of the climate change and rising sea levels on the Carterets" (Rakova 2014, 274). Following the Bishop's visit, a partnership between Tulele Peisa and the Catholic Church was established.

Tinputz became the first relocation site. Here the Catholic Diocese gifted 71 hectares of church land (Edwards 2013, 66).[3] In 2007 the Carterets/Tinputz

Relocation Task Force Committee was established, comprising representatives of Carteret Islanders, Tulele Peisa, the Catholic church and the host community. Community profiles were elaborated, counselling sessions conducted and community meetings and awareness raising on the relocation from the Carterets to Tinputz carried out. In April 2009 the first wave of families from the Carterets arrived on Bougainville, the heads of five families and their sons. They were to pave the way for the others (around 100 members of extended families). They had to work hard clearing the bush and making gardens. Adjusting to the conditions in the new environment was difficult, and there were also arguments with local landowners. Three of the initial settler families became homesick or distressed and returned to the Carterets.[4] They were replaced by others who were prepared to carry out physical work (Rakova 2014). Each family is allocated one hectare of land for food crops, cash crops and family housing. By 2019, ten families with 103 family members lived in Woroav village, at the Tinputz relocation site. Due to limited space, not more than ten families can be accommodated here. The Woroav relocation site is a pilot project, and Tulele Peisa has initiated this as a learning site to be replicated in Tearouki, Wakunai, Tenapo and Keriaka.

The resettlement plan does not only address issues such as constructing housing and infrastructure for the settlers, but also envisages the implementation of agricultural and income generating projects (mainly around cash cropping of coconuts and cocoa) and to support and help improve existing facilities such as health and education in the host communities, the development of additional education and health facilities as well as community development training programmes which will support the settlers in adjusting to their new home environment economically and socially (Tulele Peisa, n.d.). A series of workshops and trainings has been conducted over the last years, addressing issues such as food security, forestry rehabilitation and land use management (Rakova 2014, 282).

The plan also takes into account the needs and interests of the target communities so as to "ensure that these host communities will also benefit through upgrading of basic health and education facilities and training programs for income generation" (Tulele Peisa, n.d., 5). The reason for this is to avoid preferential treatment of relocated newcomers because this could cause resentment, frustration and animosities from the side of host communities. Tulele Peisa has put a lot of reflection and effort into this problem, trying to establish sustainable bonds between newcomers and recipient communities, and developing inclusive programmes which are of benefit for both settlers and hosts. Particular attention has to be paid to equity issues so as to avoid situations in which newcomers are better (or worse) off than the members of host communities, as this can easily spark resentments and conflicts. For example, Tulele Peisa is very anxious to make sure that settlers do not end up with bigger and better houses than their Bougainvillean neighbours. Moreover, the resettlement plan envisages "exchange programs involving chiefs, women and youth from host communities and the Carterets (…) for establishing relationships and understanding" (Rakova 2009, 2). Several such programmes have been actually carried through: chiefs and elders exchanges and learning visits between the Carterets and the resettlement site in Tinputz, in Tearouki and Manetai, and Young People Environmental Speaking tours. In the speaking tours, Carterets youth "meet with their counterparts in Bougainville to discuss the context of their relocation, the pressures of climate change, and the ways they can work together to make the most of the relocation reality" (UNDP 2016, 9).

Tulele Peisa deliberately promotes intermarriages as a means of relationship-building, with intermarriage seen as "a bonding cultural tie which is also a tool for preventing conflicts, resentments and jealousies" (Rakova 2014, 276). It is,

> a peace initiative and an avenue of building lasting relations between the relocated families or community with the host community of Tinputz (…) when the relocated families marry their sons and daughters into the host community, it further builds confidence and a sense of truly belonging as this signifies human interaction and connectedness.
>
> (Rakova 2014, 285).

While some settlers agree with this approach, others are opposed to intermarriages, arguing that intermarriages will be destructive for the maintenance of one's own culture (Lange 2009, 90).

Resettlement was accompanied by custom ceremonies which farewelled people on the Carterets and welcomed them to host communities on Bougainville. This included the exchange of shell money, pigs and food with the Catholic Diocese and the landowning clans in Tinputz (the Naruen and Nakaripa, Naboin, Natasi and Nakas) (Rakova 2014, 287). As both the Carteret Islanders and the Bougainvilleans are Melanesians, they share a common cultural background which makes building relationships and mutual understanding relatively easy. Furthermore, clans on the Carterets have long-standing kinship ties to clans on the mainland. Things will be more difficult for people from the other atolls in the ARoB (Mortlocks, Tasman, Nuguria) as they are Polynesians who do not have such kinship ties with people on mainland Bougainville or the adjacent island of Buka.

Substantial work has been completed or is underway in the Tinputz resettlement site: clearing of the site, establishing food gardens (for own consumption and sending surplus back to the Carterets), planting taro, learning how to grow cocoa trees and ferment and dry cocoa beans for income generation, building houses, planting trees, setting up water tanks. Houses were built by the settlers themselves, with the support of carpenters and workers from the host communities and with local building materials. This provides a new source of income for host communities, and it also "creates a sense of ownership and pride for the local community to contribute and showcase their tradesmanship" (Rakova 2014, 281). Each home has a 9,000-litre rainwater tank. Stored water carried the relocatees "through a longer-than-normal two-month dry season" in 2018 (Caritas 2018, 33). By now, the gardens which have been established by the settlers "produce enough food for relocated islanders to meet their subsistence needs, to earn a living from selling cash crops, and to send back food to relatives remaining in the Carterets" (UNDP 2016, 10). Tulele Peisa runs a 'Mini Food Forest' project: more than 34,000 trees including hard and soft wood, fruit and nut trees and five varieties of palm trees, with cassava, bananas and swamp taros planted inside the forest. The plan is to replicate this initiative at the three other relocation sites. Tulele Peisa also runs a small agricultural research station (holding over 20 different species of yams), and it owns two and a half hectares of land with over 2,500 hybrid cocoa and 500 cocoa clone trees which provide some cash income for the organisation. Cocoa farming is envisioned as the future basis of sustainable livelihoods for resettled Carteret Islanders. Tulele Peisa established Bougainville Cocoa Net in 2009, a collective business enterprise, as an income generation device for relocated families as well as local cocoa producers from the host communities (Pascoe 2015, 81). Fifty farmers are directly involved in the Tulele Peisa cocoa project, and around 1,500 farmers are members of Bougainville Cocoa Net. So far, Bougainville Cocoa Net has rehabilitated 8,950 hybrid cocoa trees over 14 hectares in Woroav and 50 hectares of cocoa plantations around the Iris Village Assembly area in Tinputz; it exports cocoa to overseas buyers, for example, in Hamburg, Germany. Tulele Peisa is committed to sustainable natural resource management at the relocation sites (Rakova 2014).

Tulele Peisa is governed by a seven-member Board of Directors, two of whom are the Chairmen of the local governing bodies in the Carterets and in Tinputz respectively. This placement of the Chairmen on the Board is "a key mechanism to ensure that local voices are represented in the leadership of the organization" (UNDP 2016, 7). Tulele Peisa currently has six people on its staff, led by the Executive Director (Ursula Rakova). Other staff members are the Program Manager who is also female, the Relocation and Community Development Co-ordinator (female), the Climate Change & Adaptation Co-ordinator (male), the Finance & Administration Manager (female) and the Fundraising & Marketing Co-ordinator (male).

The Land Problem

Securing more land for the people who want to resettle will be the most important and most difficult issue. Tulele Peisa is planning to have four more resettlement sites in addition to Tinputz, namely Tearouki, Wakunai, Tenapo and Keriaka. Land is scarce on Bougainville, and traditional land tenure in Bougainville societies does not easily lend itself to accommodation of newcomers. The vast majority of land on Bougainville (95 percent) is covered by customary land tenure (Pascoe 2015, 78). Only small portions are alienated land which, at some stage in colonial times, was bought or expropriated by outsiders—churches, white plantation owners or the colonial administration. It is no wonder that the settlers from the Carterets were relocated to land in the possession of the Catholic Church. This land—71 hectares in Tinputz, 295 hectares altogether at the four relocation sites—is by far not enough. According to the resettlement criteria developed by Tulele Peisa, some 1,500 hectares of land will be required to accommodate all the 1,500 families who intend to resettle (one hectare of land for every relocated family). It will be extremely difficult to negotiate the acquisition of customary land between Carteret Islanders and communities on Bougainville and to obtain clear legal title to land. Respective negotiations with landholders in resettlement sites started in 2007 and are continuing. Securing the funds for land purchase is another critical issue. Tulele Peisa estimates that 14 million Kina (approximately 5.3 million USD) is needed to relocate all of the families who wish to move (UNDP 2016, 13). Hence, "financial constraints remain a major challenge for the relocation efforts" (Pascoe 2015, 81).[5]

Members of the Carterets Relocation Task Force Committee are in continuous dialogue about the thorny land issue with the elders, chiefs and church leaders of the host communities. Preparations for relocation to Tearouki (which is in the vicinity of Woroav village, the Tinputz relocation site) were furthest advanced, with the establishment of a Tearouki Relocation Committee in November 2013 and community resource mapping and land surveys underway. However, local Tearouki settlers from inland communities have since moved into the Church plantation land claiming to rent the land which had been set aside to be settled by the Carteret Islanders in Tearouki. Currently discussions are being held with the Bougainville Lands & Physical Planning Department on the issuance and clarification of the land title to enable the lawyers to continue in the drafting of the land deed to be signed between the Catholic Church and Tulele Peisa.

Getting access to land on mainland Bougainville is not the only land-related problem. Leaving their own land behind is also problematic for Carteret islanders. In fact, to have to leave their own land is a shocking prospect. They are afraid of losing their cultural heritage, their identity and dignity which are closely linked to the land (Edwards 2013, 67; Pascoe 2015, 76). People "have a special, emotional bond to the tsitsiki, or land. Owning land gives islanders an identity and a sense of belonging" (UNDP 2016, 5). Abandoning one's land and thus the ancestors is a traumatic experience. Chief Paul Maeka from Han island in the Carterets explains: "The hardest thing will be to lose our sacred places, our tambu places" (quoted from Pacific Institute of Public Policy 2009, 2). This is why there are still people who do not want to leave (Connell 2018, 87), and why some people who do move suffer from mental health problems (Edwards 2013, 74). It is particularly the elderly who do not want to move, while members of the younger generation are more willing to do so with the outlook of building a sustainable economic base for their young families.

More practical issues matter too. Atoll islanders have problems adapting to other environments (Edwards 2013, 70). They have to learn new agricultural techniques, getting used to other food crops and to planting cash crops, to living detached from the sea, further inland or even on high ground, without the fishing opportunities they grew up with. For relocated Carteret Islanders in Tinputz, for example, it is a major problem that they now live a little bit far away (several minutes' walk) from the beach on a hillside, and they cannot go fishing because they do not have a boat, and they have to adjust to growing and eating unfamiliar food (Edwards 2013, 74). Therefore, a risk "that comes with displacement and relocation is that traditional knowledge is lost or becomes irrelevant in new surroundings, creating feelings of isolation and, more pragmatically, threatening the potential of relocated people to provide for their own material needs" (UNDP 2016, 5).

Resettlement poses particular challenges for women. On the Carterets and in most parts of Bougainville, communities are matrilineal, which means that land is transferred from the mothers to their daughters. The loss of land is a traumatic experience for the Carterets women as the chain of land transfer will be broken. On the other hand, the women realise that their land cannot sustain the families any longer. They are torn between the desire to stay and the need to move if they want to secure a future for their children.

Tulele Peisa is trying to take these emotional and psychological factors into account as far as possible. In Woroav the one hectare of land is in the hands of the mothers of the families, trauma counselling is provided, and the resettlement plan envisages the establishment of a regular sea transport service for freight and passengers in order to maintain links between relocated people and those who will stay behind. The establishment of a Conservation and Marine Management Area around the Carterets is also envisaged so as to maintain the area as customary fishing ground and thus keep the links to the ancestral land (Tulele Peisa, n.d., 6). Currently, Tulele Peisa is advocating for the Carterets to be declared a marine protected area (MPA) under national law of Locally Managed Marine Areas (LMMA) (UNDP 2016, 11).

The fears and concerns of the Carteret Islanders are a strong reminder that resettlement is not only a technical issue concerning mainly material problems, but also has a highly important cultural, psychological and even spiritual dimension. Tulele Peisa is fully aware of this dimension; one of its objectives is to "assist Carterets people to overcome fear, anxiety and trauma associated with the need to leave their homeland" (Tulele Peisa, n.d., 8).

External Support (Or the Lack of It)

The plight of the Carteret Islanders has drawn considerable international attention. They were presented as being at 'the frontline of global climate change' and dubbed the world's first 'environmental refugees', and their relocation was presented as "one of the first organised resettlement movements of forced climate change migrants anywhere in the world" (Displacement Solutions 2008, 2). In fact, the Carteret Islands achieved "iconic status" in the international public discourse on climate change and migration (Connell 2018, 73). More than two dozen film crews, news networks and freelance media teams have visited the Carterets over the last few years and have spread the Carterets message to the outside world (critical on this: Connell 2018). In fact, so many media people have visited that they have become a burden and locals have at times banned them from entering the islands. Representatives of Tulele Peisa have been on speaking tours to Australia, New Zealand, North America, and Europe including England and Germany. "International climate conferences, seminars, and workshops have put the group and the plight of the Carterets people on the radar of international climate scholars, policymakers, and activists" (UNDP 2016, 12). So far, however, all this international public attention has not translated into substantial support or benefit for the Carteret Islanders. The current resettlement programme which is conducted by Tulele Peisa is dependent on the resources and ingenuity of the Carteret Islanders themselves, plus modest support from donors and international civil society especially the Embassy of Finland in Canberra and church organisations in Germany, the United States and Australia.

Tulele Peisa was successful in building linkages with international civil society and other indigenous communities around the world which are affected by climate change similar to the Carterets people. Partnerships exist, for example with the international network of Climate Wise Women, with Friends of the Earth Australia, Oxfam New Zealand, Australian Conservation Foundation, Caritas New Zealand and Caritas Australia, the Christensen Fund, the Newtok Community in Alaska and the Ahus Community on Manus Island in PNG. Native Americans from Newtok and fellow citizens from Ahus even visited Tulele Peisa, Woroav and the Carterets.[6] This,

> visit by the Alaskans and Manus Islanders was an eye opener for the visitors as well as the Carteret Islanders as they learnt that there are other people in the world who are suffering the same fate of impacts of climate change.
> (Rakova 2014, 287).

Support from the side of the state of PNG and the ABG so far has been very modest. State institutions acknowledge the problem and the need for action, but things at their end move very slowly. In October 2007 the PNG government allocated two million Kina (800,000 USD) for an official 'Carterets Relocation Program'. But none of this money was given to the Tulele Peisa programme. Instead, the ABG set up its own state-led relocation programme. So far, an office in charge of relocation has been set up by the ABG in the ABG administration centre of Buka, and the ABG adopted an 'Atoll Integrated Development Policy'(AIDP) in 2007 and formed a multi-sectoral 'AIDP Steering Committee' (Lange 2009, v). This means that the ABG is not only planning for the relocation of Carteret Islanders, but also the inhabitants of the other atolls in the ARoB (Mortlocks, Tasman, Nuguria). In 2009, after lengthy consultations with local landowners, resettlement land was secured on Buka (Karoola Plantation, a former copra plantation of 600 hectares),[7] and an 'AIDP Ground Committee' was formed with the participation of representatives from local communities (Lange 2009, v–vi). The plan was to resettle 40–60 families at the Karoola plantation.[8] In the following years, several rounds of surveys and social impact studies of the Carteret Islanders and of host communities were conducted (Edwards 2013, 64), asking atoll islanders about their concerns, needs and aspirations regarding resettlement. Workshops and Focus Group discussions were held, interviews carried out, expectations raised, but so far, no actual resettlement in the context of the state programme has taken place. It is not clear how much of the money has been used already for preparatory work, and how much is left for actual resettlement. Funding for actual relocation still seems to be a major problem.[9] More recent initiatives like the establishment of a Bougainville Climate Change Office and the launch of the 'Building Resilience for Climate Change for Bougainville' programme as part of the PNG-wide national Climate Change Program in May 2018 so far have not led to actual progress either.[10]

Unfortunately, relations between Tulele Peisa and the ABG are difficult. The ABG attacked Tulele Peisa for running a parallel relocation programme to the ABG's programme and refused to support Tulele Peisa. Relationships between some Carteret Islanders who occupy high-ranking positions in the ABG and the Bougainville public service on the one hand and the leadership of Tulele Peisa on the other are strained. Carteret Islanders reconciliation and unification ceremonies which took place in Buka and on the islands from June 2018 onwards are supposed to initiate a 'Carterets unification process' so as to overcome divisions in the community and between leaders.[11] Such unification would be highly welcome in the interest of the people of the Carteret Islands. A major step forward in this regard was made in 2018. On 23 November 2018, a major reconciliation involving more than 2,700 islanders, and witnessed by some World Bank representatives, ABG politicians, including the then Vice President who is also from the Carterets, and senior Carterets public servants, took place on the Carterets. This reconciliation addressed all the outstanding cases from the time of war. It was a full success. It also established collaborative partnerships between the ABG Administration, the Carterets community and its local level government, and Tulele Peisa. This is in recognition and appreciation of each other's efforts and for the betterment of Carteret Islanders as a whole. Tulele Peisa today is working together with some ABG departments, especially the Lands Department. Tulele Peisa has also been asked to be involved in elaborating the ABG's climate change policy.

Expectations and Concerns of Relocated People

From the surveys and community consultations conducted in the context of the ABG's AIDP, a number of important issues regarding relocation governance become obvious.[12] People have high hopes and expectations, but they also voice serious concerns and fears, and they have very specific demands and proposals regarding the planning process and the implementation of relocation.

People expect from relocation: food security, better access to services, in particular health and education, better opportunities for income generation (jobs, cash crops) and useful exchanges of skills and knowledge with host communities.

People are concerned about: the loss of traditional lifestyles (based on close ties with the sea) and traditional culture (due to abandonment of the land of their ancestors and interaction with host communities and an 'alien' society), loss of local language and changes in religious practices. They fear for their safety due to land disputes and conflicts with host communities (including fear of sorcery); there is also fear of changes in the status and role of women and youth and fear of an increase in alcohol and drug consumption due to the availability of cash. The Polynesian relocatees additionally are afraid of racism by the Melanesian majority on Bougainville.

People have long lists of demands regarding planning and implementation of relocation, including: the continuous involvement of communities via regular consultations with community leaders/chiefs; establishment of all essential infrastructure, services and facilities at the resettlement site before actual relocation (schools, churches, health posts, police stations etc.); permanent housing reflecting the traditional village layout; allocation of sufficient land for agriculture; improved transport between atolls and mainland; separate sites for the different island communities so that they can maintain their culture and feel safe; security provision; guarantees for the maintenance of the clan system of islanders (resettlement of entire clans in one site) and of their traditional leadership system.

Concerns of (potential) relocatees very much revolve around the question of how relationships with host communities will play out: will they be hostile or friendly? Anxieties abound (Lange 2009, 139), and experiences of relocatees often confirm concerns. Some of the settlers from the Carterets were re-relocated to their home islands because of bad experiences with their neighbours. Most difficult are the cases where relocatees have to negotiate access to customary land, and even if the resettlement land is formally legally free (so called alienated freehold land) and thus in principle available for resettlement, in most cases there are people already there, dwelling there, gardening there, or hunting—'illegally' perhaps according to state law, but with reference to long-established customary rights of usage. This is the case with Karoola plantation land on Buka. Karoola plantation is legally freehold land, but nevertheless the ABG had to negotiate access with the neighbouring communities whose members have used this land for a long time.

Planned relocation can lead to local conflicts between settlers and recipient communities. This was the case in two previous state-led Carterets-Bougainville resettlement endeavours from the early 1980s and the late 1990s which both failed because of such conflicts (Edwards 2013, 63–64). The first, the Kuveria programme, called the Atolls Resettlement Scheme, commenced in 1982. The plan was to resettle a total of 40 families. The resettlement site at Kuveria on Bougainville, however, turned out to be inappropriate; it was adjacent to the provincial correctional facility, and with an unwelcoming host community that denied the resettled families fishing rights and access to land. Only 15 families moved to Kuveria. Soon, "differences and feuds often flared up between the settlers and the traditional landholders" (Rakova 2014, 275). Consequently, after a few years "the resettlement attempt failed and the majority of settlers returned to the Carterets" (UNDP 2016, 6).

The second state-led endeavour had a similar fate. In 1997 the Bougainville Administration relocated 30 Carterets families to the Hanahan area in the northeast of Buka island. But most of them "returned to the Carterets as a result of land disputes (…) there was a failure to integrate the new arrivals into the receiving community and the government withdrew its support for the resettled families after the initial relocation period" (UNDP 2016, 6). Relocatees complained about "a lack of 'unity' with the host community" (Lange 2009, 103), with ongoing conflicts over land use and fishing rights. Relocatees were the target of hostilities from their neighbours who destroyed their houses and food gardens or their produce when they took it to the market or attacked their young people or raped the women (Lange 2009, 104). As a consequence, "many families returned to the Carteret Islands due to difficulties integrating with the host community" (Lange 2009, 104).

UNDP summarises the causes of the failure of these unsuccessful resettlement schemes as follows: "Reluctance on the part of relocating communities to move, a lack of local voices in the design and execution of resettlement—including location choice—lack of suitable land, and insufficient attention to social integration with host communities" (UNDP 2016, 6).

Conclusion

The Carteret Islanders' case illustrates that it is difficult to make a clear distinction between 'forced' and 'voluntary' instances of migration related to climate change and its effects. Tulele Peisa stresses the point that it organises 'voluntary' resettlement. On the other hand, people feel 'forced' off their homeland. They would not leave if the atolls were not becoming uninhabitable. "The line between voluntary migration and forced displacement from climate change (…) is difficult to determine" (UNDP 2016, 4). Instead of getting too occupied with debates about the 'voluntary/forced' problem—and the legal implications that come with it—it therefore might be more advisable to focus on the opportunities for long-term planned resettlement in cases where it can be predicted that people will have to move sooner or later. Planning—and implementation of plans—should start early so that people have the opportunity to relocate voluntarily in a well organised manner sooner, rather than being forced into a hasty, disorganised move later. To make this option a reality is exactly what Tulele Peisa is trying to do.

Long-term planning is a must—and it is possible.

> When it comes to sea-level rise in particular, there is no need to wait for extreme weather events to strike and islands and coastal regions to be flooded. All areas that cannot be protected through increased coastal defences for practical or economic reasons need to be included early in long-term resettlement and reintegration programmes that make the process acceptable for the affected people.
>
> (Biermann and Boas 2010, 83).

As the Carterets case demonstrates, in order to make processes 'acceptable for the affected people', comprehensive community participation is vital. In fact, at "the core of Tulele Peisa's work is the active promotion of community self-reliance and ecologically and culturally sustainable relocation" (Pascoe 2015, 78). This is why UNDP is of the view that Tulele Peisa's "community-based approach to relocation offers a positive relocation model for other atolls in the region" (UNDP 2016, 3).

The Carteret Islanders have not waited for the state and others to come to their assistance, but have taken their fate into their own hands, and in doing so have shown considerable capabilities and ingenuity. The people on the ground have agency of their own; they are not just passive victims of climate change. On the other hand, local agency should not be used as an excuse for inaction of international or regional organisations and of governments and state institutions (Pascoe 2015, 85). One should not let international organisations and national governments off the hook.

In the Carterets case, it is obvious that there are problems in the relations between state institutions and communities. The state has not delivered yet, and Carteret Islanders have taken things into their own hands, so far largely detached from state institutions. On the side of the state there are serious governance deficiencies, as the failure of state-run attempts at relocation demonstrates.

In a post-conflict situation like in Bougainville, states are often relatively weak and fragile; they have problems functioning effectively and delivering services to their people. In such fragile situations, it is of particular importance that state institutions and international donors work closely together with non-state actors: civil society actors like NGOs and community-based organisations, but also customary local networks and traditional authorities. They are in charge of the governance of communities, natural resources and the environment; they have to be involved in climate change adaptation, including resettlement measures. Resilience of communities and adaptive capacity very much rest with these customary actors and institutions and the indigenous traditional knowledge of which they are custodians (Boege 2018). They can and do play an important role in planning, decision-making and implementation of resettlement programmes, as the example of Tulele Peisa demonstrates. The interesting thing about Tulele Peisa is that it is not just an NGO or civil society organisation in the Western understanding of the term, but is closely linked to non-state actors who do not fit neatly into the Western 'civil society' category. Tulele Peisa was set up at the request of the local Council of Elders, that is, traditional authorities from the customary sphere of societal life. Tulele Peisa thus can be seen as an example of a "bridging organisation" (Petzold and Ratter 2015, 40), which connects local customary life-worlds and the 'outside' world of state and civil society.

Flowing from these conclusions, a few recommendations can be drawn in order to increase the chances of success of the Carterets—Bougainville relocation. These are:

– Enhance efforts for Carterets reconciliation, in particular including Tulele Peisa and the ABG (responsible: ABG, Tulele Peisa, Carterets communities and their leaders, churches);
– Build linkages between the ABG resettlement programme and the Tulele Peisa programme so that they become mutually supportive (instead of just being pursued in parallel or even in competition) (ABG and Tulele Peisa, with third party mediation and facilitation);
– Support the implementation of the Carterets Integrated Relocation Plan, in particular funding for more housing (government, donors, INGOs);
– Expand and support the economic activities of Bougainville Cocoa Net Limited and the Carteret Islands Investments (CII)[13] (Tulele Peisa, cocoa farmers, ABG, donors);
– Negotiate access to additional land for Carteret Islanders willing to resettle, and secure formal land titles (Tulele Peisa, ABG, Catholic Church, host communities);
– Establish regular and reliable transport to and from the Carterets (ABG);
– Establish a marine protected area (MPA) under national law for the Carterets (PNG government, ABG).
– Pay special attention to the most important aspects that have made Tulele Peisa significant in piloting the CIRP. These are: recognising the rights of the relocated families to land security and land use management, their rights to safe homes and secure property ownership (of family blocks or plantations) so as to maintain their dignity as human beings and communities also after relocation.

Notes

1 The ARoB comprises the main island of Bougainville, the major island of Buka, and several small islands and atolls. Geographically, Bougainville is part of the Solomon Islands archipelago in the Southwest Pacific. It is approximately 9000 sq km in size (the size of Cyprus) and has approximately 300,000 inhabitants.
2 It is estimated that around 300 Carteret Islanders live elsewhere in PNG, mainly in Buka Town, the administrative centre of the ARoB (Connell 2018, 81–82).
3 The church intends to give formal land rights to the settlers for 30 hectares. So far this has not happened, hence rights to the land are not guaranteed (Edwards 2013, 68). But Tulele Peisa is currently working together with a law firm on a land deed to transfer the leases to the families, under the women's family names, because according to Carterets custom it is the women who hold the custodianship of the land.

4 Post Courier Wednesday 22 July 2009. Islanders return to Bougainville atolls. By Gorethy Kenneth.
5 A proposal for a significant expansion of the Tulele Peisa relocation project was handed in to New Zealand's Ministry of Foreign Affairs and Trade in June 2018. Unfortunately, it was unsuccessful in a competitive funding round (Caritas 2018, 52). Tulele Peisa is continuing to seek funding for its programme from external donors.
6 The National, Monday 10th September 2012. Alaskans to assist people of Carteret.
7 There are serious problems with Karoola plantation land: it is contested, it is swampy, difficult to access by road, and with difficult access to the sea.
8 New Dawn on Bougainville 03/09/13. Resettlement of Carteret Islanders soon to start. By Aloysius Laukai. In 2013 it was announced that relocation to Karoola plantation was imminent, but "relocation to Karoola Plantation in 2014 did not eventuate because of no funding". Gorethy Kenneth, Post-Courier Tuesday, 8 December 2015.
9 New Dawn on Bougainville 29/12/17. Climate change funding slow. By Aloysius Laukai.
10 New Dawn on Bougainville 16/05/18. Climate Change affecting all islands.
11 New Dawn on Bougainville 03/06/18. Carteret Unification Effort acknowledged.
12 For the following see Lange 2009, in particular the summary 139–142.
13 CII is the coordinating body for all business and economic activities on the Carterets.

References

Biermann, Frank, and Ingrid Boas. 2010. "Preparing for a Warmer World: Towards a Global Governance System to Protect Climate Refugees." *Global Environmental Politics* 10 (1): 60–88.

Boege, Volker. 2018. "Climate Change and Conflict in Oceania: Challenges, Responses, and Suggestions for a Policy-Relevant Research Agenda." Toda Peace Institute Policy Brief No 17. Tokyo: Toda Peace Institute.

Campbell, John R. 2014. "Climate-Change Migration in the Pacific." *The Contemporary Pacific* 26 (1): 1–28.

Caritas Oceania. 2018. "Waters of Life, Oceans of Mercy: Caritas State of the Environment for Oceania 2018 Report." Wellington: Caritas Aotearoa New Zealand.

Connell, John. 2018. "6 Nothing There Atoll? 'Farewell to the Carteret Islands.'" In *Pacific Climate Cultures: Living Climate Change in Oceania,* edited by Tony Crook, and Peter Rudiak-Gould, 73–87. Warsaw and Berlin: De Gruyter.

Displacement Solutions. 2008. "The Bougainville Resettlement Initiative." Meeting Report. Canberra: Displacement Solutions.

Edwards, Julia B. 2013. "The Logistics of Climate-Induced Resettlement: Lessons from the Carteret Islands, Papua New Guinea." *Refugee Survey Quarterly* 32 (3), 52–78.

Lange, Kirstin. 2009. *Bougainville Atolls Resettlement. An Exploration of Key Social and Cultural Impacts.* Buka: Bougainville Administration Division of Community Development.

Pacific Institute of Public Policy. 2009. "Climate Countdown: Time to Address the Pacific's Development Challenges." Port Vila: Pacific Institute of Public Policy (Briefing 12).

Pascoe, Sophie. 2015. "Sailing the Waves on our own: Climate change migration, self-determination and the Carteret Islands." *QUT Law Review* 15 (2): 72–85.

Petzold, Jan, and Beate M.W. Ratter. 2015. "Climate Change Adaptation under a Social Capital Approach – An analytical Framework for Small Islands." *Ocean & Coastal Management* 112: 36–43.

Rakova, Ursula. 2009. "How to Guide for Environmental Refugees." http://ourworld. unu.edu/en/how-to-guide-for-environmental-refugess/ (accessed 12 October 2010).

Rakova, Ursula. 2014. "The Sinking Carterets Islands: Leading Change in Climate Change Adaptation and Resilience in Bougainville, Papua New Guinea." In *Land Solutions for Climate Displacement*, edited by Scott Leckie, 268–290. London and New York: Routledge.

Tulele Peisa Inc, n.d. "Carterets Integrated Relocation Program." Bougainville, Papua New Guinea. Project Proposal. Buka: Tulele Peisa.

United Nations Development Programme. 2016. "Tulele Peisa, Papua New Guinea." Equator Initiative Case Study Series. New York.

7 Changing Environments and Conflict Prevention in Solomon Islands

Kate Higgins

Introduction

Indigenous knowledges and ways of being in Pacific societies, Vaai explains, are not compartmentalised into neat analytical boxes, such as 'nature', 'economy', 'human endeavour', 'spirituality', 'politics' and so on, but rather are interdependent aspects of lived experience forming part of a relational whole (2019; also Bhagwan 2019). Local conceptions and practices which aim to produce peace and justice are located, not only within individuals or institutions, but also in the significantly powerful 'spaces between' (see Vaai 2019, 8). Connection with one's environment is a key aspect of the relational worlds in which peace and security are governed in Solomon Islands. Peace and justice are embedded in beliefs, practices, processes and institutions woven into landscapes and seascapes. The ocean and islands which make up the physical environment of Solomon Islands are changing. This is due to a combination of factors such as population growth and natural resource (mis)management. The changing climate is also a key factor which is impacting upon the relationship between people and their environments. Therefore, while there is no direct correlation between conflict and climate change, climate change must be considered as a factor in the overall Solomon Islands peace and conflict context.

Solomon Islands is an archipelago made up of almost 1,000 islands in the South-Western Pacific Ocean. Around 80 percent of the estimated population of 680,000 people live in small-scale rural settlements (World Bank 2018; Solomon Islands National Statistics Office 2019), although Solomon Islands is experiencing rapid rates of urbanisation (Keen et al. 2017, 13; Campbell 2019b). Solomon Islands experienced low-level civil conflict between 1998 and 2003 which was driven, among other factors, by historical patterns of internal migration, the natural resource management and distribution, and the dissatisfaction felt among citizens towards the ill-fitting model of the centralised postcolonial state (Bennett 2002; Braithwaite et al. 2010; TRC 2012). Despite histories of conflict, it is important to recognise the considerable adaptive capacity of Solomon Islanders, as demonstrated by continued survival over centuries of colonialism, capitalism, missionisation, war, conflict and—important to note here—extreme weather events. Communities

DOI: 10.4324/9781003001744-10

continue to organise their own ways of living in interaction with the institutions surrounding them, mixing indigenous and introduced practices resulting in *relative* peace and stability at a community level (see Boege et al. 2008). In supporting peace and livelihoods, working with existing adaptive capacity is a key challenge facing the state and external actors as life in Solomon Islands continues to change.

The changing climate is a significant challenge which the island nation faces. Solomon Islands is experiencing higher temperatures, fluctuations in rainfall, and more frequent El Nino weather patterns (Birk and Rasmussen 2014, 2). Sea surface temperatures are increasing, and ocean acidification and rising sea levels are contributing to declining fish stocks through the destruction of coastal habitats and reefs (Dey et al. 2016, 1–2). Increased soil salinity and erosion from rising sea levels affects food gardens (Asugeni et al. 2017, 1–2). Climate change is intensifying natural disasters, such as floods and cyclones, as well as weather patterns which cause prolonged droughts and heat waves (Birk and Rasmussen 2014, 2). In some locales, such as the artificial islands[1] and coral atolls, sea level rise, king tides and food and water security are impacting upon settlement itself and resulting in internal migration (Monson and Fitzpatrick 2016). Stress on the physical environment is therefore likely to impact upon existing community capacities to manage complex social relations which are orientated around governing land and resources—social relations which have emerged over long periods of time through interaction with the natural environment.

Indigenous governance and justice mechanisms which prevent and respond to conflict are often legitimised by collectively remaking place-based identities, identities which are embedded in the relationship between humans, non-humans and their environments. In rural areas, where the most land is customarily held, communities draw upon a mixture of indigenous and introduced and adapted governance processes to maintain peaceful social relations and relations around land and resources. These forms of indigenous governance constitute the primary mode of conflict prevention in operation across the archipelago. When disputes arise, indigenous justice mechanisms administrated by chiefs, elders and church leaders (sometimes in conjunction with police, often not) are key in restoring *relative* peace (McDougall and Kere 2011; Allen et al. 2013). The effectiveness of state institutions (such as police, schools and health clinics) are largely dependent on the participation and support of local leaders. Moreover, in urban locales such as Honiara, indigenous modes of conflict prevention and resolution "travel"; while these modes adapt to new circumstances in the process, indigenous leaders continue to mediate between state institutions and the "travelling traditions" of migrants in urban areas (Boege and Hunt 2020).

Given the relational and embedded nature of peace and justice in Solomon Islands, this chapter examines the nexus between changing physical environments and the capacity of Solomon Islanders to prevent violent conflict. Specifically, this chapter discusses conflict prevention amidst changing environments in relation to three factors impacting upon social and political order. First, that climate change is a "threat multiplier" following this prevalent idea found in understandings of climate change and conflict (Boege 2018, 6). In this section, I examine existing sources of conflict in Solomon Islands in relation to ongoing environmental degradation. Second, the relationship between internal migration issues linked to climate change and conflict. Internal migration has historically been a conflict driver in Solomon Islands. Now there are cases of communities needing to relocate as well as increasing rates of urbanisation which present a risk to peace and stability. Third, I explore indigenous, state and internationally led responses to climate change at community level. In particular, I argue that there are potential conflict risks associated with externally designed projects such as climate change adaptation or disaster risk reduction interventions. Overall, this chapter seeks to highlight key questions: how do governments, civil society actors, policymakers, and external actors assist with the prevention and mitigation of conflict impacted by continued environmental change? How do these actors work alongside communities to ensure conflict sensitive adaptation? How can we draw upon local adaptive capacities rather than create dependencies while avoiding the common forms of conflict associated with outside forms of intervention in order to contribute to meaningful adaptation?

Climate Change as a Conflict 'Multiplier' in Solomon Islands?

It would be a fallacy to make direct links between current conflict drivers in Solomon Islands and the impacts of climate change (see Boege 2018, 6). However, it is also a mistake to ignore how rapidly changing physical environments in Solomon Islands will impact upon peace and stability within community life. What happens when water becomes scarcer? Or if the land available for growing food and cash crops is reduced? What happens when the increased salinity of soil reduces the quality of agriculture land? What happens to the health of people as proteins found in fish become harder to acquire (Albert et al. 2015; Dey et al. 2016)? Finally, what happens to the social, spiritual, psychological and relational worlds of communities—connected to landscapes and seascapes—as environments change around them? These are not easy questions to answer. However, what is now clear is that climate change must be understood as a critical factor impacting upon social and political order. One approach is to examine the ways in which environmental change might exacerbate existing conflict occurring across the country.

The foundation of many conflict issues is land and land disputes are often cited as the most significant form of conflict in communities (Allen et al. 2013, 18–19). As in other contexts in Oceania (Campbell 2019a), land questions are highly complex in Solomon Islands. Land should not be understood only in economic terms, nor as the physical location in which people live, but rather must also be understood in terms of its social, relational and cosmological or spiritual dimensions. Local cosmological worldviews are crucial to people's identity and sense of wellbeing and consequently their ability to participate in, and willingness to maintain, the 'social harmony' upon which local under-standings of peace are constructed. In this context, 'cosmology' refers to the ways in which people understand the universe they inhabit—its origins and futures—and incorporates stories of ancestors, narratives of the coming of Christianity, and place-based histories which retell of the people who have come and gone, leaving their mark on the landscape and seascape (Ballard 2014). Land underpins a sense of identity and belonging which links present generations to the ancestors of the past while safeguarding it for those yet to come (Ballard 2014; see also Kempf et al. 2014). One's home organises one's social relations (that is, one's '*wantok*'[2]), not only within a village or island, but also within the nation-state and beyond (see Bolton 2003, 70). How these so-cial structures—and the power-relations embedded within them—are negoti-ated and governed is thus crucial in the capacity to prevent, mitigate or resolve conflict within rapidly changing physical environments (Boege 2018, 8).

Eighty-seven percent of land in Solomon Islands is customarily held by kin-ship groups (McDonnell et al. 2015, 13). Customary land arrangements differ significantly across the country. Arrangements are generated and maintained through diverse practices including through lineage structures, marriage and adoption practices, and customary forms of transfers and payments (ibid.). Conflict over land increases when economic benefits are at stake (Allen et al. 2013, 18). Improved resource management—especially as resources become scarcer—is a key aspect of conflict prevention. Most rural settlements are semi-subsistent, relying on small-scale agriculture for food production, as well as on cash crops such as copra and cocoa. The increase in the need for cash for school fees, transport, and imported goods such as rice, alcohol and medicines has led to the development of cash-cropping over many decades, disrupting cycles of intergenerational land inheritance (ibid., 18–19). There are pushes to register land through state-legal mechanisms, with the hopes this will cre-ate conditions for economic development. However, registering land is not necessarily a prerequisite to economic activity (McDonnell et al. 2015) and widespread land and resource disputes occur on both 'registered' and 'unreg-istered' land. Local justice and governance mechanisms which seek to produce peaceful relations—regardless of whether they are 'statutory' or 'customary', or, as is often the case, a mix of both—are often orientated around maintain-ing land relations. With land at the centre of many intersecting conflict issues, the potential loss of the useable land through environmental degradation may exacerbate existing forms of conflict.

Natural resource management is a key conflict challenge Solomon Islands faces. The high levels of resource extraction by foreign companies, most of which is both legally and ethically questionable, is a significant driver of con-flict across Solomon Islands. At the national level, the unsustainable forestry industry has generated economic and political challenges to the stability of the Solomon Islands state for decades. Logging is deeply entwined in na-tional political dynamics and underlies many of the common complaints from citizens about the pervasiveness of national corruption (see Bennett 2002; Kabutaulaka 2006; 1998; Allen 2011). In addition, a small mining industry is growing. Given the forms of conflict associated with logging, as well as lessons learnt from the serious conflict experiences generated by mining in neighbouring Bougainville, there is significant potential for mining to increase the likelihood of conflict in Solomon Islands (Porter and Allen 2015; Allen 2017; 2018). Attempts to regulate these industries have had little impact, and in fact, Foukona and Timmer explain how the state acts as a "capitalist landlord", treating customary land as "estates" and is facilitating the very transactions which cause conflict (2016; see also McDougall 2016, 221–222). Nationally regulated sustainable resource management is extremely unlikely to occur in the short to medium term, conflating existing environmental degradation.

Logging and mining often result in local social conflicts, causing "lasting rifts between and within landholding groups, villages, families and households", while also reinforcing "gender inequity by systematically excluding women from decision-making and from sharing in the benefits" (Minter et al. 2018, 6). The effect of extractive industries on gender relations as well as influxes of outside labour (both foreign workers and workers from other parts of Solomon Islands) has been cited as a factor affecting the high levels of gender-based violence the nation experiences (SPC 2009, 10). Influxes of cash as a result of logging and mining is feeding endemic substance abuse issues with substance abuse often cited by community members as a key and common source of instability in communities (Allen et al. 2013, 27–30). Extractive industries not only disrupt social order by creating jealousies and conflict over who benefits, logging and mining also leave destructive marks on the environment, impacting upon food security by damaging forests, gardens, mangroves and reefs as well as polluting water sources (see Minter et al. 2018, 6). The combination of environmental impacts stemming from climate change, and environmental and social effects of resource mismanagement may be exacerbating the stress that communities are under, and impact upon capacities to prevent or manage conflict around land and resources within communities.

Insecurity in Solomon Islands is also attributed to the mixed and changing demography (Allen et al. 2013). Not only is the Solomon Islands a culturally and linguistically diverse nation characterised by small-scale group identities, the population of Solomon Islands is growing and is expected to surpass one million by 2050 (UNFPA 2014, 70, 72). This will add stress to existing land and sea resources, a factor which may undermine stability. Demographic data shows that the population of Solomon Islands is young, with 60% of people under the age of 25 (ibid., 72). Intergenerational tension is a commonly expressed community grievance while the legitimacy of decision-making by chiefs and elders is under stress (Allen et al. 2013, 16–18). This is perceived as impacting upon the effectiveness of existing localised conflict prevention and resolution mechanisms which tend to be dominated by elders (ibid.).

Conflict is not only localised in Solomon Islands as there is significant dissatisfaction at the nature of the centralised state for failing to recognise, support and deliver to rural communities. This was a major factor in the low-level civil conflict which occurred from 1998 until an Australian-led international intervention *The Regional Assistance Mission to Solomon Islands* (RAMSI) put an end to fighting in 2003 (see Braithwaite et al. 2010; Fraenkel et al. 2014). While the conflict has been stabilised, many of its underlying causes remain unresolved. These include land issues, historic and current internal-migration issues, uneven development, the management of the logging industry, and the anger of citizens at the nature of the centralised state and the continuous failures of decentralisation policy (TRC 2012; Dinnen and Allen 2016). While the state, and its associated 'formal' justice mechanisms (police and courts), is idealised as the key mediating entity between the different interests driving conflict (Cudworth et al. 2007, 3), state institutions as they currently exist, including actors who manipulate or mimic state institutions, often cause or exacerbate conflict both at national and local levels. As McDougall explains, "[c]itizens may have much to gain from a better government, but they also have much to lose from the strengthening of a state they have good historical reason to mistrust" (2016, 222). The centralised state, and the actors who draw upon state power, are the most significant potential drivers of conflict in Solomon Islands. This is important considering the tendency, Monson and Fitzpatrick observe, "to emphasise state-based norms and institutions... in shaping adaptation to climate change" (2016, 240). I elaborate upon this problem further in sections on migration and adaptation below.

Finally, there are questions around how changing environments impacts upon cycles of intergenerational trauma. It was reported to one researcher that "Solomon Islanders may have order but they do not have 'peace in their hearts'" (George 2018, 1324). Cycles of intergenerational trauma stem, not only from the civil conflict and the present post-conflict environment, but also from deeper histories of violence. These may include the extraordinarily disempowering form in which colonialism took place in Solomon Islands (see Bennett 1987), the impact of the Second World War, and an array of intergenerational conflict and violence which has occurred in different locales and which is embedded within the worlds of communities and families. Conflict legacies may have bearing on the forms of social disorder occurring across the country. In cases where climatic and environmental change adds further forms of stresses, or in the worst of cases results in displacement, it is important to consider the social and psychological impact of climate change. This has been described by some as "ecological grief" (Cunsolo and Ellis 2018) and has obvious links to community resilience and adaption. However, on this point it is also worth exercising caution: there is also a need to avoid the tendency present in some academic, policy and activist circles to dramatise and sensationalise the 'issue' of climate change—of playing on emotions and co-opting 'their' situation into 'our' existential crisis. This could lead to just the latest form of negative external representations of Pacific Islanders as small, disconnected and vulnerable (see Hau'ofa 1994).

Migration and Conflict

As with the relationship between climate change and the conflict context discussed above, there is no clear-cut correlation between climate change and internal migration in Solomon Islands; in fact in cases where climate change-affected populations are relocating there are many social, economic and political "push and pull" factors (Birk and Rasmussen 2014, 2). However, the relationship between internal migration and climate change is a key factor in examining the nexus between climate change and conflict. Internal migration has long been a feature of life in Solomon Islands, and there is frequent circular movement to and from villages to provincial centres and/or to the capital, Honiara. Moreover, there are long histories which tell of the movement and settlement of people. This is often a strong feature of local oral histories which relate how a group of people have arrived at a certain place (see for example, Scott 2000). While connection to one's *'home'*—meaning one's own island and land—is ubiquitous in local discourses of peace and security, migration and settlement is often also a significant feature of place-based histories which detail the relationships between people and their environment (see Monson and Fitzpatrick 2016). Nonetheless, for low-lying islands and atolls, the risk of losing one's home is a real threat, one that is likely to cause political, social, economic, spiritual and psychological uncertainty and distress. Climate change-related migration is already occurring—although often *in combination* with other social, economic and political factors—and much of this migration is taking the form of urban drift to Honiara.

For sake of ease, climate change-related migration can be understood in different categories: 'institutionally-led', 'community-led', and 'family or individually-led', although all three are interrelated and complex. Institutionally led relocation involves a scenario where outside institutions including governments or churches assist with relocations of people due to climate change impacts. Based on instances to date, this approach remains fraught. Institutionally led relocation hinges on the historically difficult interaction between state and customary systems and highlights the lack of legitimacy of the centralised state, particularly in relation to land governance (Monson and Fitzpatrick 2016). There are examples of attempts of institutionally led relocation. First is the case of the provincial capital of Choiseul, Taro. The provincial government has plans to move the provincial capital from the low-lying island of Taro to the mainland. The Taro relocation "is the first time that a provincial capital with all its services and facilities will be relocated in the Pacific Islands" (Scientific American 2014). However, it is likely this relocation will take time to achieve as the Solomon Island government is now looking for the support of international donors in order to implement the relocation plan (Haines 2016).

The case of the low-lying atolls of Ontong Java is a more complex example for it is not a government station like Taro; rather is made up of Polynesian customary groups. Talk of relocation is made difficult not only due to internal disagreement about whether people are willing to relocate (Solo, n.d.), but also by confusion around the slow and complex nature of the attempts of the provincial government, as well as the Anglican church, to negotiate a new place to which people can relocate (Fr. Nigel Kelaepa, personal communication, Honiara, 30 August 2018). Reportedly, the Autonomous Region of Bougainville in neighbouring Papua New Guinea has been suggested as a site, as has the island of Malaita which is part of the same province as Ontong Java (ibid.; Monson and Fitzpatrick 2016, 252). In either case, there is potential for conflict. In Bougainville, this is evidenced by the difficulties experienced by Carteret Islanders who moved from their atolls to mainland Bougainville (Boege and Rakova 2019). In the second case, some community members report that they fear inter-group conflict over land will emerge if they are to move to Malaita (Fr. Nigel Kelaepa, personal communication, Honiara, 30 August 2018; Solo, n.d.). Instead people from Ontong Java have said they would prefer to migrate over time through intermarriage into other island groups (ibid; Monson and Fitzpatrick 2016, 252). Meanwhile, the people of Ontong Java are experiencing significant food security issues associated with salinity, while rising seas have already destroyed buildings and reduced land for settlement. Many members of the Ontong Java community have migrated to a settlement in Honiara, located at the mouth of the Mataniko River (Birk and Rasmussen 2014, 8). Overall, attempts by the government to facilitate relocations are likely to be fraught for both those who are relocating as well as the recipient communities. Given current state capacities, especially in relation to land governance, institutionally led migration is unlikely to be a widely adopted any time soon.

The second type of climate change-related migration, community-led relocation, involves a negotiation between different customary groups occurring outside of state-based mechanisms. Community-led relocation is not an option which can be initiated top-down by state or external actors. As Monson and Fitzpatrick explain, this type of relocation "…take[s] place according to terms, concepts, and cultural frameworks provided by customary land systems…embedded in historical networks of intermarriage, kinship, trade and exchange" (2016, 247–248). This option is highly dependent on the capacity of local leadership, the availability of land, and the historical nature of social relations within an area. For example, the community of Lilisiana in Malaita Province has negotiated relocation through customary networks (ibid., 250–251). Likewise, after becoming continually inundated by king tides and rising seas, Walande, an artificial island in the South of Malaita Province negotiated with customary landowners on the mainland for land to settle (ibid; Higgins and Maesua 2019).

Once settlement has been negotiated there is also conflict risk over 'rights' to resources for cultivating food and income, as occurred in the case of the Walande settlers. While disputes over resource rights were reconciled using customary justice mechanisms, many settlers have now chosen to move to urban centres due to problems with the settlement site itself. It was found to be muddy and filled with flies and mosquitoes which cause illness, and it was also found to be difficult to adjust to the change in culture required to go from living on the 'saltwater', to moving into the 'bush'[3] (Higgins and Maesua 2019). It is also important to consider intergenerational issues, as other experiences of customary land arrangements and disputes have demonstrated that what has been negotiated within and between kin groups in one generation, may not hold the same legitimacy in future generations (McDonnell et al. 2015). While there are significant challenges, customary negotiations are a crucial adaptive strategy, a strategy which builds upon the existing adaptability of Solomon Islanders, and the centrality of kinship networks in the production of peace and livelihoods.

Family or individually led relocation is the most common type of climate change-related migration. As Birk and Rasmussen argue "In the context of uncertain climate futures… migration at the levels of individuals and households represents an important adaptive strategy" (2014, 12). Care must be taken in these cases to make direct causal links as there are a variety of 'push and pull' factors which drive people to move from their villages to provincial centres or Honiara (ibid.). However, individuals and families are moving from areas where there are food and water security issues, reduced land for economic activity, and population growth. People are seeking livelihood opportunities elsewhere, including in provincial centres; such is the case of Reef Islanders who have relocated to Lata, Temotu Province (Birk and Rasmussen 2014), and those from atolls and artificial islands who have relocated to Honiara (Higgins and Maesua 2019). Moreover, many of those who move are often young men who move for work or education, often ending up in informal settlements that suffer from poor access to basic services, poor housing quality and unemployment.

Solomon Islands has some of the highest rates of urbanisation in the Pacific at 4.33 percent (Campbell 2019b, 3; see also Keen et al. 2017, 13). The urbanisation rate is approximately twice that of Papua New Guinea (2.2 percent) and Fiji (1.7 percent) and if this rate continues, the population of Honiara will double in only 16 years (Campbell 2019b, 3). Migration is enabled by freedom of movement to Honiara, and as Foukona and Allen explain, the historical precedents which have resulted in the emergence of settlements within and beyond Honiara town boundaries (2017, 88–92). There are also historical precedents for migratory-related conflict; while many cases of historical migration across the Solomon Islands have been relatively peaceful, migration and resettlement are also a historic conflict driver. This is evidenced in the violent conflict which occurred on Guadalcanal, where fighting between the indigenous Guale and migrants escalated from 1998 (Moore 2007). Furthermore, informal and 'illegal' settlements in Honiara remain vulnerable to climate change impacts and natural disasters. They are often built on sites which are at greater risk of storm surges (Keen and McNeil 2016; Ha'apio et al. 2018). While current violence within Honiara tends to result from political issues associated with anger at the government (discussed further below), looking longer term, an increased population with limited space for settlement could present risks for peace and stability within the capital.

Local, State and International Climate Adaptation

Processes and practices of governance within communities are not static, nor are people unable to adapt to changing environments. Communities have coped with climate variability and extreme weather events over centuries, successfully maintaining levels of well-being in highly uncertain environments (Monson and Fitzpatrick 2016, 248; Warrick et al. 2016, 1042). Despite the existence of both government policy and significant interest in climate change adaptation and resilience from the international community, questions remain as to how rural and urban communities can best be assisted to meet uncertainty related to climate change and to potential conflict impacts now and into the future. Attention and financial support are not enough as processes and relations required must be given serious thought. The prevention and mitigation of conflict is particularly challenging given current ambiguities in the relationship between the small-scale communities scattered throughout the island archipelago and highly centralised state institutions (White 2007; Monson and Fitzpatrick 2016). The relationship (or lack of) between small-scale communities and external development actors who tend to be based in Honiara or elsewhere, is another factor to consider in whether forms of social and environmental adaptation may also create forms of conflict.

Local Adaptation

Evidence of how communities are adapting to climate change impacts is "highly context specific" (Warrick et al. 2016, 1048). A myriad of local power-relations and factors determine the potential and actual capacities of communities to adapt to climate change impacts, and to do so with conflict sensitivity. Warrick et al. observe that global understandings of adaptative capacity focus on strong economies, strong states, and technology transfer as key adaptive factors; however, in Solomon Islands, navigating adaptation within existing social and political relations is the key requisite for success (Warrick et al. 2016, 1041). A significant political factor in a community's ability to adapt is local leadership. For example, community leaders in Pileni, low-lying islands in the Reef Islands of Temotu Province, have played an important role in decision-making processes, conflict mediation, and cultural forms of reconciliation which have enabled community members to adapt in uncertain circumstances (ibid., 1046). Asugeni et al. describe the local innovations in adaptation efforts of the villages in East Kwaio in Malaita province, bringing together indigenous leadership practices and knowledge of the natural environment as well as leadership from influential community members working in the health centre to combat sea level rise (2017). Albert et al. describe how indigenous leadership has helped to maintain fish catches, despite evidence of fish stocks reducing due to climatic change, by changing the methods and/or the locations in which community members fish (2015; see also Sulu 2010). Leadership capacity does not depend only on the capabilities of the leader, but also the historical and present-day political formation of communities (see Stasch 2010) and the complex web of relations in which the leader is located, is required to navigate, and at times "untangle" (see Brigg 2008, 6).

Another key factor impacting upon adaptive capacity is the way the powerful force of Christianity is practised in different settings. There are differing understandings of climate change among different Christian denominations. As with other locales (Vaai 2019, 11), some see climate change impacts such as rising seas or king tides as a result of sin or as a "curse" (One New Limited, n.d. cited in Higgins and Maesua 2019, 11). For example, Ha'apio et al. (2018) describe how church leaders in one village in Western Province believe they are protected from climate change given that their village is a historically important Christian site (361–362). Conversely, the church in Ontong Java (discussed above) is actively speaking about climate change, running adaptation programmes, and making attempts to assist with national conversations about the need to relocate (Solo, n.d; Fr. Nigel Kelaepa, personal communication, Honiara, 30 August 2018). This mirrors broader attempts within the Pacific to 'decolonise theology' and work towards articulating a climate change-relevant theology (Vaai 2019, 11–12). Several established Solomon Islands churches are members of the Pacific Conference of Churches (PCC), a prominent civil society advocacy body in the Pacific. The PCC has declared "climate change is real" and is actively seeking to support churches to support relocating communities (PCC 2019). These are in themselves positive and conflict sensitive aspects of adaptation, especially given the influence of Christianity upon cosmological beliefs—including beliefs about where the future lies.

How community members understand climate change should not be taken for granted. Monson and Fitzpatrick observe that people living by the coast and on atolls and artificial islands are proactive in identifying observed weather phenomena as symptomatic of climate change, particularly changes in winds, tides and extreme weather conditions (2016, 242). However, knowledge about current and future climate change impacts is highly variable across the country and within each community (Albert et al. 2015; Ensor et al. 2018; Ha'apio et al. 2018; Solo, n.d). While the academic and policy worlds understand the problem to be a technical one, explained by scientific evidence, it is likely that many community members comprehend climate change in a more relational sense, where, as noted above, the "environment" and "human community" are part of a whole (Bhagwan 2019; Vaai 2019). Therefore, conflict sensitive strategies that aim to address issues of climate change adaptation must not only incorporate but also respect and value community perceptions of environmental change and engage in dialogue rather than solely 'explaining science', as can often be the case in externally led secular projects.

National Adaptation

Solomon Islands government policy has recognised the challenge that climate change poses. Climate policy understands climate change as a threat to existing resilience and as a barrier to development. Climate seeks "a resilient, secure and sustainable Solomon Islands responding to climate change" (MECDM 2012, 13). The policy lists adaptation, disaster risk reduction and mitigation capacity as key in achieving "increased resilience" and "sustainable development" goals (ibid.). There are also examples of adaptation efforts, as was noted above in the case of the Taro relocation, and plans to improve agriculture and food security, water security and sanitation, health and education in response to Climate Change (MECDM 2012, 10). There are limitations in what the government can do given its resources without external assistance. Furthermore, in terms of relocation and as discussed above, the state does not have the legitimacy to make decisions over customary land. Attempts to interfere are likely to cause conflict. The lack of legitimacy and trust in the state is also likely to muddy its role in much needed urban adaptation. For example, riots instigated by people living in settlements which occurred after extreme flooding in Honiara in April 2014 were motivated by anger at the government for failing to recognise their situation. While the state is commonly assumed to be the key actor in national climate change adaption (Monson and Fitzpatrick 2016; Warrick et al. 2016), as well as in managing conflict, this assumption remains fraught in Solomon Islands given the present relationship between the state and its citizens (Brigg 2018).

International Intervention

The post-conflict Solomon Islands context has led to a flood of international intervention, often under the banner of 'development'. Development is ubiquitous in commentary within and about Solomon Islands. In Solomon Islands government policy, development is linked to adapting to climate change (MEDCM 2012, 10). However, 'development' is a poorly defined endeavour and is also a significant driver of conflict. Support to state institutions, a preferred mode of development assistance, often fails to impact upon communities due to the weak relationship between communities and state institutions. At the community level, donor, NGO, and government projects which introduce new resources can cause jealousies and create or exacerbate existing disputes between groups, conflicts which often result in project failure (Allen et al. 2013, 23–26). How might these precedents impact upon the implementation of climate change adaptation, disaster risk reduction, or other relevant environmental projects? External interveners often fail to account for the relational nature of the existing contexts—including the existing conflict dynamics within each community—in which the project seeks to operate.

While not specific to climate change adaptation or disaster risk reduction projects, there are common issues which can result in project failure and unexpected outcomes of conflict in communities. Externally led projects often fail due to short project cycle timeframes. Often there is not enough time allowed for external actors (including locally engaged staff and volunteers) to walk alongside community members throughout the project process. This also limits the ability to recognise and develop strategies which work with existing relations and capacities, and instead to apply standardised solutions as outlined in the project document, treating each community context in an ahistorical and apolitical way. This can have the effect of failing to work with indigenous forms of knowledge (Warrick et al. 2016) and may even serve to reduce local adaptive capacities by reproducing the idea that outsiders alone have the (technical) solutions to 'fix' environmental impacts. The risk is creating further dependencies—often disparagingly referred to throughout Solomon Islands as a 'hand-out mentality'.

Finally, questions remain as to whether adaptation measures have positive and sustainable outcomes, or whether methods of adaptation may in fact be forms of *maladaptation*, making matters worse over the longer-term. Given project evaluations are often conducted a short time after a project is implemented, this makes them more difficult to measure, and consequentially to hold outside technical approaches and implementing organisations to account. In cases where maladaptation occurs (for example, Fazey et al. 2011), this has the potential to further embed climate change impacts in the overall conflict context of Solomon Islands over the longer-term. Overall, there is a need to employ conflict sensitive strategies in local, national and international adaptation efforts. At the same time, and given the increasing need for climate change adaptation, there may also be opportunities to strengthen more inclusive local governance and justice mechanisms which prevent and respond to conflict as part as a central aspect of adaptation itself.

Conclusion: Conflict Prevention Amidst Environmental Change

Given there is no direct correlation between conflict and climate change, this chapter has examined the climate change and conflict nexus through three elements of the Solomon Islands peace and conflict context. First, I explored this nexus in relation to environmental change, and in particular how environmental degradation may exacerbate conflict. Second, I explored this nexus through ongoing migration patterns, discussed the possibilities and difficulties of, as well as highlighting risks associated with, increased urbanisation. Third, and perhaps somewhat counterintuitive to some, I argued that forms of external intervention (including by state institutions) which seek to address climate change impacts could potentially constitute a form of climate change-related conflict. Existing local adaptation efforts emphasise the management of conflict dynamics, the importance of indigenous leadership, and the ongoing negotiation of social, political, economic and spiritual forms of power, adaptive elements which are often largely neglected in external approaches. In the remainder of this chapter, I conclude with some suggestions for addressing both structural and relational implications raised here.

In relation to environmental degradation and existing forms of conflict, there are no easy solutions. However, a key conflict prevention strategy is to address resource mismanagement, and in particular extractive industries. Extractive industries are enhancing land and social conflict as well as environmental destruction. The 'ideological' drive to develop economically (Filer et al. 2017, 29–31; see also Bhagwan 2019; Vaai 2019), and the politicised nature of logging and mining, continues to reproduce a national political paralysis when it comes to addressing extractive industries. While it is very unlikely that state-based environmental protections will become a reality in the short to medium-term, resource extraction must be both decreased and effectively regulated. Smaller efforts at environmental conservation and adaptation must be accompanied by larger structural change. In building peace and justice, how might climate change provide the impetus for this structural change, as impacts continue to increase? Could climate change constitute the opportunity to address destructive economic structures which drive different forms of national and local conflict and livelihood insecurity?

Climate change impacts are more severe in some locations than others, and for some communities, relocation is now necessary. This chapter drew upon examples from Ontong Java, the Reef Islands, and artificial islands of Malaita where informal migration is taking place. However, there are other cases that exist across the archipelago (MECDM 2012, 40). There is a need to better map and predict climate change-related resettlement areas, and to address both intra-group and inter-group relationships among both those relocating and existing cultural groups. In cases where relocation occurs and in the growing urban settlements in Honiara, this increased human mobility may provide an opportunity to examine and work with strategies which promote greater interconnectedness between different groups. Supporting dialogue around different indigenous governance and justice practices will assist with approaches to resolving conflict across difference. This can build upon the already considerable capacity of Solomon Islands to do so (McDougall and Kere 2011, 142–143). Given the current anger at the government present among citizens (Evans, n.d.; Kabutaulaka 2008; Allen 2013), including those who over time have relocated to informal settlements in Honiara, violent conflict cannot be prevented, mitigated or resolved by state institutions alone. In fostering peace and stability within the capital, there is an opportunity to work with the relations between different actors (Brigg 2016) including indigenous leaders who emerge in urban sites, carrying with them "travelling traditions" (Boege and Hunt 2020).

Shifting the focus to state-community relations rather than seeing the state and its development partners as delivering 'adaptation' solutions is key to managing the ongoing impacts of internal migration, as it is to managing overall adaptation to climate change. At a practical level this includes restructuring the 'project approach' ubiquitous throughout Solomon Islands. Investment in relationships between local, national and international actors and the consequential co-production of solutions will increases likelihood of success.

Allowing longer timeframes for intervention and slowing down actions to allow space for conflict sensitivity, mitigation and resolution will be necessary for success. Ensuring greater inclusion—not only of women and youth—but also inclusive representation of different kinship groups is essential. There are also common pitfalls which can be avoided; for example, rather than establishing new 'climate change' or 'adaptation' committees it is better to work with existing embedded and legitimate institutions such as churches, chiefly leadership, elders and other customary forms of leadership, women, youth representatives, and local service providers (such as teachers and health workers). Ensuring meaningful adaptation is a significant challenge given the entrenched bureaucratic processes associated with development assistance. However, as the need for climate change adaption increases, there may be opportunities to find ways to decolonise the 'project' as well as the dominance of interveners. This will ensure that the interconnected set of challenges experienced by those living together in changing environments are supported in ways which do not undermine indigenous frameworks, but rather build upon them as a foundation for conflict prevention and secure livelihoods.

Notes

1 Historically, the artificial islands of Malaita have been built in the lagoons surrounding the coast out of coral rocks and other local materials (Moore 2017, 216). They are said to have been built so as to avoid malaria-affected areas on the mainland and due to the relatively large size of the population (ibid., 21, 48).
2 *Wantok* or literally 'one talk' is a term which denotes belonging within a social structure, commonly of the same vernacular language group, of which there are approximately 80 in Solomon Islands (Monson and Fitzpatrick 2016, 241). However, the term is applied in different ways depending on how close/far one is situated from one's home. In a village *wantok* might refer to a close relative, at an island level to the vernacular language group, at the national level to an island group, and when overseas, to someone who is also from Solomon Islands.
3 A common identity differentiator across the Solomon Islands between *saltwater people* and *bush people* holds significant meaning. Bennett compares this to the difference between country and city people in Western places (1987, 6).

References

Albert, Simon, Shankar Aswani, Paul L. Fisher, and Joelle Albert. 2015. "Keeping Food on the Table: Human Responses and Changing Coastal Fisheries in Solomon Islands." *PLOS One* 10 (7): 1–13.
Allen, Matthew. 2011. "Long-Term Engagement: The Future of the Regional Assistance Mission to Solomon Islands." *Australian Strategic Policy Institute*. Strategic Insights 51.
Allen, Matthew G. 2013. *Greed and Grievance: Ex-militants' Perspectives on the Conflict in Solomon Islands, 1998–2003*. Honolulu: University of Hawai'i Press.
Allen, Matthew. 2017. "Islands, Extraction and Violence: Mining and the Politics of Scale in Island Melanesia." *Political Geography* 57: 81–90.

Allen, Matthew. 2018. *Resource Extraction and Contentious States: Mining and the Politics of Scale in the Pacific Islands*. New York: Palgrave Macmillan.

Allen, Matthew, Sinclair Dinnen, Daniel Evans, and Rebecca Monson. 2013. *Justice Delivered Locally: Systems, Challenges, and Innovations in Solomon Islands*. Washington, DC: The World Bank.

Asugeni, Rowena, Michelle Redman-MacLaren, James Asugeni, Tommy Esau, Frank Timothy, Peter Massey, and David MacLaren. 2017. "A Community Builds a "Bridge": An Example of Community-led Adaptation to Sea-level Rise in East Kwaio, Solomon Islands." *Climate and Development* 11 (1): 91–96.

Ballard, Chris. 2014. "Oceanic Historicities." *The Contemporary Pacific* 26 (1): 96–124.

Bennett, Judith. 1987. *Wealth of the Solomons: A History of a Pacific Archipelago, 1800–1978*. Honolulu: University of Hawaii Press.

Bennett, Judith. 2002. "Roots of Conflict in Solomon Islands-Though Much Is Taken, Much Abides: Legacies of Tradition and Colonialism." *State, Society and Governance in Melanesia*. Discussion Paper 5. Canberra: Australia National University.

Bhagwan, James. 2019. "Climate Initiatives Must Change: The Effect of Modern "Development" Trends on the Pacific" Speech to the Papua New Guinea Council of Churches. 20 August 2019 available at https://pacificconferenceofchurches.org/f/climate-initiatives-must-change

Birk, Thomas, and Kjeld Rasmussen. 2014. "Migration from Atolls as Climate Change Adaptation: Current Practices, Barriers and Options in Solomon Islands." *Natural Resources Forum* 38: 1–13.

Boege, Volker. 2018. "Climate Change and Conflict in Oceania: Challenges, Responses and Suggestions for a Policy-Relevant Research Agenda." Toda Peace Institute Policy Brief No.17. Tokyo: Toda Peace Institute.

Boege, Volker, M. Anne Brown, Kevin P. Clements, and Anna Nolan. 2008. "States Emerging from Hybrid Political Orders: Pacific Experiences." *The Occasional Papers* 10, no. 11. Brisbane: Australian Centre for Peace and Conflict Studies.

Boege, Volker, and Charles T. Hunt. 2020. "On 'Travelling Traditions': Emplaced Security in Liberia and Vanuatu." *Cooperation and Conflict* 55 (4): 497–517.

Boege, Volker, and Ursula Rakova. 2019. "Climate Change-Induced Relocation: Problems and Achievements – the Carterets Case." Toda Peace Institute Policy Brief No.33. Tokyo: Toda Peace Institute.

Bolton, Lissant. 2003. *Unfolding the Moon: Enacting Women's Kastom in Vanuatu*. Honolulu: University of Hawai'i Press.

Braithwaite, John, Sinclair Dinnen, Matthew Allen, Valerie Braithwaite, and Hilary Charlesworth. 2010. *Pillars and Shadows: Statebuilding as Peacebuilding in Solomon Islands*. Canberra: ANU E Press.

Brigg, Morgan J. 2008. "Networked Relationality: Indigenous Insights for Integrated Peacebuilding." Hiroshima University Partnership for Peacebuilding and Social Capacity Discussion Paper Series No. 3.

Brigg, Morgan. 2016. "Relational Peacebuilding." In *Peacebuilding in Crisis: Rethinking Paradigms and Practices of Transnational Cooperation*, edited by Tobias Debiel, Thomas Held, and Ulrich Schneckener, 56–69. London and New York: Routledge.

Brigg, Morgan. 2018. "Beyond the Thrall of the State: Governance as a Relational-affective Effect in Solomon Islands" *Cooperation and Conflict* 52 (2): 154–172.

Campbell, John R. 2019a. "Climate Change, Migration and Land in Oceania." Toda Peace Institute Policy Brief No.37. Tokyo: Toda Peace Institute.

Campbell, John R. 2019b. "Climate Change and Urbanisation in Pacific Island Countries." Toda Peace Institute Policy Brief No.49. Tokyo: Toda Peace Institute.

Cudworth, Erika, Tim Hall, and John McGovern. 2007. *The Modern State: Theories and Ideologies.* Edinburgh: Edinburgh University Press.

Cunsolo, Ashlee, and Neville R. Ellis. 2018 "Ecological Grief as a Mental Health Response to Climate Change-related Loss." *Nature Climate Change* 8 (4): 275–281.

Dey, Madan Mohan, Kamal Gosh, Rowena Valmonte-Santos, Mark W. Rosegrant, and Oai Li Chen. 2016. "Economic Impact of Climate Change and Climate Change Adaptation Strategies for Fisheries Sector in Solomon Islands: Implication for Food." *Marine Policy* 67: 171–178.

Dinnen, Sinclair, and Matthew Allen. 2016. "State Absence and State Formation in Solomon Islands: Reflections on Agency, Scale and Hybridity." *Development and Change* 47 (1): 76–97.

Ensor, Jonathan Edward, Kirsten Elizabeth Abernethy, Eric Timothy Hoddy, Shankar Aswani, Simon Albert, Ismael Vaccaro, Jason Jon Benedict, and Douglas James Beare. 2018. "Variation in Perception of Environmental Change in Nine Solomon Islands Communities: Implications for Securing Fairness in Community-based Adaptation." *Regional Environmental Change* 18 (4): 1131–1143.

Fazey, Ioan, Nathalie Pettorelli, Jasper Kenter, Daniel Wagatora, and Daniel Schuett. 2011. "Maladaptive Trajectories of Change in Makira, Solomon Islands." *Global Environmental Change* 21 (4): 1275–1289.

Filer, C. Siobhan McDonnell and Matthew Allen. 2017. "Powers of Exclusion in Melanesia." In *Kastom, Property and Ideology: Land Transformations in Melanesia*, edited by Siobhan McDonnell, Matthew Allen, and Colin Filer, 1–56. Canberra: ANU Press.

Foukona, Joseph D., and Matthew G. Allen. 2017. "Urban Land in Solomon Islands: Powers of Exclusion and Counter-Exclusion." In *Kastom, Property and Ideology*, edited by Siobhan McDonnell, Matthew Allen, and Colin Filer, 85–110. Canberra: ANU Press.

Foukona, Joseph D., and Jaap Timmer. 2016. "The Culture of Agreement Making in Solomon Islands." *Oceania* 86 (2): 116–131.

Fraenkel, Jon, Joni Madraiwiwi, and Henry Okole. 2014. "The Ramsi Decade: A Review of the Regional Assistance Mission to Solomon Islands, 2003–2013." Independent Report.

George, Nicole. 2018. "Liberal-local Peacebuilding in Solomon Islands and Bougainville: Advancing a Gender-just Peace?" *International Affairs* 94 (6):1329–1348.

Ha'apio, Michael Otara, Morgan Wairu, Ricardo Gonzalez, and Keith Morrison. 2018. "Transformation of Rural Communities: Lessons from a Local Self-initiative for Building Resilience in the Solomon Islands." *Local Environment* 23 (3): 352–365.

Haines, Philip. 2016. "Choiseul Bay Township Adaptation and Relocation Program, Choiseul Province, Solomon Islands." *Case Study for CoastAdapt, National Climate Change Adaptation Research Facility.* Gold Coast, Australia. Accessed 27/02/19 online at https://coastadapt.com.au/sites/default/files/case_studies/CSS3_Relocation_in_the_Solomon_Islands.pdf

Hau'ofa, Epeli. 1994. "Our Sea of Islands." *The Contemporary Pacific* 6 (1):148–161.

Higgins, Kate, and Josiah Maesua. 2019. "Climate Change, Conflict and Peacebuilding in Solomon Islands." Toda Peace Institute Policy Brief No.36. Tokyo: Toda Peace Institute.

Kabutaulaka, Tarcisius Tara.1998. "Deforestation and Politics in Solomon Islands." In *Governance and Reform in the South Pacific,* edited by Peter Larmour, 121–153. Canberra: National Centre for Development Studies, ANU.

Kabutaulaka, Tarcisius Tara. 2006. "Global Capital and Local Ownership in Solomon Islands' Forestry Industry." In *Globalisation and Governance in the Pacific Islands,* Studies in State and Society in the Pacific No. 1, edited by Stewart Firth, 239–258. Canberra: ANU E Press.

Kabutaulaka, Tarcisius Tara. 2008. "Westminster Meets Solomons in the Honiara Riots." In *Politics and State Building in Solomon Islands,* edited by Sinclair Dinnen, and Stewart Firth, 96–118. Canberra: ANU EPress.

Keen, Meg, Julien Barbara, Jessica Carpenter, Daniel Evans, and Joseph Foukona. 2017. "Urban Development in Honiara Harnessing Opportunities, Embracing Change." Research Report. Canberra: ANU.

Keen, Meg and Alan McNeil. 2016. "After the Floods: Urban Displacement, Lessons from Solomon Islands." In Brief series. Canberra: ANU.

Kempf, Wolfgang, Toon van Meijl, and Elfriede Hermann. 2014. "Movement, Place-Making and Cultural Identification: Multiplicities of Belonging." In *Belonging in Oceania: Movement, Place-Making and Multiple Identifications,* edited by Elfriede Hermann, Wolfgang Kempf and Toon van Meijl, 1–24. New York: Berghahn Books.

McDonnell, Siobhan, Joseph Fukuona and Alice Pollard. 2015. "Building a Pathway for Successful Land Reform in Solomon Islands." State, Society and Governance in Melanesia Report. Canberra: ANU.

McDougall, Debra. 2016. *Engaging with Strangers: Love and Violence in the Rural Solomon Islands.* New York: Berghahn Books.

McDougall, Debra, and Joy Kere. 2011. "Christianity, Custom, and Law: Conflict and Peacemaking in the Post-conflict Solomon Islands." In *Mediating across Difference: Oceanic and Asian Approaches to Conflict Resolution,* edited by Morgan Brigg, and Roland Bleiker, 141–162. Honolulu: University of Hawai'i Press.

Ministry of Environment, Climate Change, Disaster Management and Meteorology (MECDM). 2012. *National Climate Change Policy 2012–2017.* Policy Report. Honiara, Solomon Islands.

Minter, Tessa, Grace Orirana, Delvene Boso, and Jan van der Ploeg. 2018. "From Happy Hour to Hungry Hour: Logging, Fisheries and Food Security in Malaita, Solomon Islands." Penang, Malaysia: World Fish Program Report 7.

Monson, Rebecca, and Daniel Fitzpatrick. 2016. "Negotiating Relocation in a Weak State: Land Tenure and Adaptation to Sea-level Rise in Solomon Islands." In *Global Implications of Development, Disasters and Climate Change: Responses to Displacement,* edited by Susana Price, and Jane Singer, 240–255. Asia Pacific: Earthscan.

Moore, Clive. 2007. "The Misappropriation of Malaitan Labour: Historical Origins of the Recent Solomon Islands Crisis." *Journal of Pacific History* 42 (2): 211–232.

Moore, Clive. 2017. *Making Mala: Malaita in Solomon Islands 1970 – 1930.* Canberra: ANU Press.

Pacific Conference of Churches. 2019. "Climate Change Is Real" website, https://pacificconferenceofchurches.org/climate-change-is-real

Porter, Doug, and Matthew G. Allen. 2015. "The Political Economy of the Transition from Logging to Mining in Solomon Islands" *State, Society and Governance in Melanesia.* Discussion Paper 12. Canberra: ANU.

Secretariat of the Pacific Community. 2009. "Solomon Islands Family Health and Safety Study: A Study on Violence against Women and Children." Report. Honiara: SPC.

Scientific American. 2014. "Township in Solomon Islands is 1st in Pacific to Relocate due to Climate Change." August 15, 2014. Accessed online 27/02/19 at https://www.scientificamerican.com/article/township-in-solomon-islands-is-1st-in-pacific-to-relocate-due-to-climate-change/

Scott, Michael W. 2000. "Ignorance Is Cosmos; Knowledge Is Chaos: Articulating a Cosmological Polarity in the Solomon Islands." *Social Analysis: The International Journal of Social and Cultural Practice* 44 (2): 56–83.

Solo, Rex. n.d. "The Plight of Resettling Atoll Communities in the Solomon Islands: The Case of Ontong Java/Lord Howe Islanders." Report. Suva, Fiji: Pacific Conference of Churches.

Solomon Islands National Statistics Office. 2019. "Population: Projected Population by Province 2010–2025" Accessed online 01/09/2019 at https://www.statistics.gov.sb/statistics/social-statistics/population.

Solomon Islands Truth and Reconciliation Commission. 2012. *Confronting the Truth for a Better Solomon Islands: Final Report*. Honiara, TRC.

Stasch, Rupert. 2010. "The Category 'Village' in Melanesian Social Worlds: Some Theoretical and Methodological Possibilities." *Paideuma Mitteilungen zur Kulturkunde* 56: 41–62.

Sulu, Reuben John. 2010. "Multidisciplinary Appraisal of the Effectiveness of Customary Marine Tenure for Coral Reef Finfish Fisheries Management in Nggela (Solomon Islands)". PhD Thesis, Newcastle University.

United Nations Population Fund Pacific, 2014 "Population and Development Profiles: Pacific Island Countries." https://pacific.unfpa.org/sites/default/files/pub-pdf/web–140414_UNFPAPopulationandDevelopmentProfiles-PacificSub-RegionExtendedv1LRv2_0.pdf

Vaai, Upolu Luma. 2019. "We Are Therefore We Live: Pacific Eco-Relational Spirituality and Changing the Climate Change Story." Toda Peace Institute Policy Brief No.56. Tokyo: Toda Peace Institute.

Warrick, Olivia, William Aalbersberg, Patrina Dumaru, and Rebecca McNaught. 2016. "The 'Pacific Adaptive Capacity Analysis Framework': Guiding the Assessment of Adaptive Capacity in Pacific Island Communities." *Regional Environmental Change* 17 (4): 1039–1051.

White, Geoffrey M. 2007. "Indigenous Governance in Melanesia." *State, Society and Governance in Melanesia*. Discussion Paper 5. Canberra: ANU.

World Bank. 2018. "Total Population" [Website]. Accessed online 12/12/2018 at https://data.worldbank.org/indicator/SP.POP.TOTL

8 Social Implications of Climate Change in Vanuatu

Potential for Conflict, Avenues for Conflict Prevention, and Peacebuilding

Kirsten Davies

Introduction

The Intergovernmental Panel on Climate Change (IPCC) declared that "Climate-related risks to health, livelihoods, food security, water supply, human security, and economic growth are projected to increase with global warming of 1.5°C and increase further with 2°C" (IPCC 2018). Scientists and policy makers have long recognised that the impacts of climate change can threaten community stability and accelerate conflict. For instance, in 2014 the United States (US) *Climate Change Adaptation Roadmap* identified climate change as potentially adding "to the challenges of global instability, hunger, poverty, and conflict" (DoD 2014, 1). The report stated, "Food and water shortages, pandemic disease, disputes over refugees and resources, more severe natural disasters – all place additional burdens on economies, societies, and institutions around the world" (ibid.) Indeed, the former United Nations (UN) Secretary-General Ban Ki-moon stated, "the scarcity of food and water [will] transform peaceful competition into violence…and droughts spark massive human rights migrations, polarising societies and weakening the ability of countries to resolve conflicts peacefully" (United Nations 2017b). The UN Deputy Secretary-General, Amina Mohammed, supported this position when she said,

> the impacts of climate change go well beyond the strictly environmental. Climate change is inextricably linked to some of the most pressing security challenges of our time. It is no coincidence that countries that are most vulnerable to climate change are often the most vulnerable to conflict and fragility.
>
> (United Nations 2018)

As part of the South Pacific region, the Republic of Vanuatu (Vanuatu) endures the brunt of the most severe impacts of climate change. The UN has categorised Vanuatu as a Least Developed Country (Garschagen et al. 2016, 6), a determination based on factors including the nation's rapid population growth (United Nations 2010, 7), coupled with food and water insecurity,

DOI: 10.4324/9781003001744-11

economic health challenges, changing climatic patterns and the increasing frequency and severity of extreme climatic events (Schaar 2018, 15). Vanuatu is ranked first on the *World Risk Index*, which is calculated by the *UN Institute for Environment and Human Security* according to natural disaster risk (Garschagen et al. 2016, 46).

Vanuatu is an archipelago of 82 islands, most of which are inhabited by small rural communities, living in coastal areas. Of Vanuatu's estimated population of 271,000 people, 70 percent are coastal dwellers. Additionally, most islands in the archipelago have an elevation of merely 0.9m (United Nations 2017c). As such, the citizens of Vanuatu (known as 'Ni-Vanuatu') are increasingly vulnerable to extreme coastal weather events. Climatic events can create a chain of extended indirect and wide-reaching consequences. For instance, saltwater inundation can result in coastal erosion, which impacts the viability of a village's location and infrastructure, and in the worst case, rendering an island uninhabitable resulting in forced relocation. This disconnection from homelands can be a catalyst for substantial social upheaval and possible conflict.

Previously known as the New Hebrides, Vanuatu has not been immune to conflict throughout history. Tribal conflicts, colonisation by France and England, and fights for independence have all plagued the nation's history. Additionally, many nations in the South Pacific, including Vanuatu, played a major role in the Second World War. Historic remnants, such as shipwrecks and ruined buildings, continue to serve as visible reminders of battles won and lost. For example, the town of Luganville, on the Vanuatu island of Espiritu Santo, was home to a major US military force of more than 500,000 people during the height of the Second World War (ESTA 2018).

Despite its tranquil image Vanuatu is susceptible to, and can be overcome by, conflict. When the New Hebrides was governed by an Anglo-French Condominium, tensions erupted between the citizens of Vanuatu and colonisers (Gubb 1994, 3). Collective community action (MacQueen 1980, 235) resulted in the nation gaining its Independence on 30 July 1980 (ibid.). Collective community action is now rising again against the adverse effects of climate change.

The Ni-Vanuatu people have an intricate and multi-faceted relationship with the environment, relying on subsistence or part-subsistence livelihoods. Most depend on the land and sea for food and water security. Additionally, nature embodies the heart of tribal affiliations and social structures that include matrilineal and patrilineal inheritance systems (United Nations 2017c). The islands hold exceptional cultural and linguistic diversity, including 108 living languages, which is more per unit area than any other country. This includes deep traditional ecological knowledge and elaborate customary laws and practices, known as *kastom* (Kwa 2008, 5). The traditional systems of governance have been managed through a system of chiefdoms, which has been challenged by the introduction of Western law and religion. As a consequence, law, theology and governance each operate from a paradigm of hybridisation, which inevitably affects the management of communities and the environment.

This chapter examines the social implications of climate change in Vanuatu, including potential drivers of conflict. It focuses on strategies for conflict prevention and peace building. It has been designed in two parts. First, it identifies aspects of Vanuatu's climate vulnerabilities, including threats to livelihood security. It explores international case studies relevant to Vanuatu's preparedness for the impacts of climate change, and climate migration options. Second, it investigates potential future pathways for Vanuatu in responding to climate change vulnerabilities. These include utilising localised approaches which demonstrate the growing power of climate change as a community connector.

Recommendations in this chapter will assist the development of domestic and international policies responding to growing evidence of the impacts of climate change on peace and security in Vanuatu.

Vanuatu's Climate Vulnerabilities

Livelihood Security in Vanuatu

Seventy five percent of the rural population in Vanuatu is responsible for more than half of the nation's agricultural production, forming a vital part of Vanuatu's food security. Vanuatu's food security is dependent on the high degree of biological diversity across the islands (Nurse et al. 2014, 4). The nation's significant reliance on natural resources is increasing because of its rapidly growing urban and rural populations (ibid.). This increased demand and consumption not only places additional pressure on the environment, but overburdens waste disposal capacities and contributes to a decline in biodiversity. The Pacific Regional Environment Program summarised Vanuatu's key climate change vulnerabilities between 2016 and 2030:

> Key challenges facing Vanuatu in the context of environmental management and development planning include rapid population growth and local population pressures; land tenure; water pollution, waste disposal and urbanization; a lack of awareness and understanding about environmental problems; depletion of key species such as coconut crabs and mangroves; inappropriate land use practices that may result in erosion and degraded soils, contributing to impacts on coral reeds and other ocean resources; invasive species; loss of forests and biodiversity; and the over-exploitation of natural resources and climate change.
>
> (SPREP 2017).

Since agriculture is one of the key sources of employment and income across Vanuatu (Buhaug 2018) the risk of violence among communities is likely to increase when drought, floods, or land overuse and degradation contribute to reduced production and economic loss (Schaar 2018, 6). Poor and vulnerable communities are more susceptible to conflict, as they do not have the same capacity to adapt (Blondel 2012, 28). For example, long-term drought may lead to a chronic breakdown in social relationships, which, in turn, may result in conflict (ibid., 10).

Changing climatic patterns in the South Pacific are challenging food and water security in Vanuatu (UNICEF 2017, 50). The availability of water for communities and agriculture is becoming increasingly uncertain, and predicted to deteriorate (ibid.). Vanuatu citizens have reported water shortages as the result of changing precipitation patterns, inadequate water infrastructure, and management and distribution challenges. The 2012 visit of the UN Special Rapporteur on the right to safe drinking water and sanitation recognised shortcomings in legal and institutional frameworks. Despite efforts to establish frameworks, finite resource and funding constraints have resulted in limited capacity for improvement (Mudaliar et al. 2015, 18).

Similarly, changing climatic patterns are challenging food security, with crop production being especially vulnerable. On land, excessive rain can stunt growth, water-log soil, or provide conditions that promote plant pests and diseases, while excessive dry seasons can reduce productivity (Reti 2007, 5). In the ocean, acidification and thermal stress are immediate threats to Vanuatu's marine ecosystems and coral reefs, which are the source of a range of traditional food. Ninety percent of Vanuatu's reef system is expected to reach a critical state by 2030 (Johnson, Welch and Fraser 2016, 11). Aquaculture is at risk of both ocean acidification and overfishing, causing marine depletion in nearshore areas (ibid., 2).

International Case Studies Informing Vanuatu's Preparedness for the Impacts of Climate Change

The Arab Spring, the Civil War in Syria, and conflict in the Lake Chad region exemplify the 'multiplier effect' of climate change, exacerbating existing vulnerabilities, leading to conflict. During the Arab Spring, extreme droughts and heat waves ruined the harvest, and the Egyptian government could no longer sustain the subsidisation of international wheat. Consequently, the price of bread tripled, activating widespread civil unrest (Sternberg 2012, 519). In Syria, drought exacerbated water and agricultural insecurity, leading to economic loss in rural areas and overall large-scale internal migration to semi-urban areas, which contributed to the outbreak of civil war in 2011 (ibid.). In 2017, the UN Security Council (UNSC) adopted a resolution on the conflict in the Lake Chad region. It explicitly identified climate change as a contributing factor to the region's instability (Schaar 2018, 7). Further, the G7 group has identified Lake Chad as a potential case for linking climate change with security threats, based on connections between drought, food insecurity and reduced livelihood options (ibid., 17).

These case studies demonstrate that the impacts of climate change can act as a threat multiplier, destabilising communities, contributing to the activation of conflict (Kelley et al. 2015, 3241). As planetary temperatures rise the likelihood of conflict increases. Climate change will exacerbate vulnerabilities of varying levels of influence. For example, there is an increased likelihood of collective violence due to the occurrence of droughts in ethnically divided societies (Brzoska 2018, 323), and natural disasters add further tension to societies where there are other vulnerabilities, such as political repression (ibid.). As Scheffran et al. explain, "the consequences of climate change depend on how vulnerable affected natural and social systems are and how sensitively they respond to the stress" (2012, 2).

Vanuatu is in the global warming 'front line'. Climatologists predict an increase in the frequency and severity of extreme climatic events (Wong et al. 2014). For example, in 2015, Cyclone Pam created damage totalling 64 percent of Vanuatu's GDP, and left 75,000 residents homeless, while destroying 96 percent of crops, placing the country's food security at risk. The cyclone destroyed infrastructure, in what was a stark foretelling of Vanuatu's future challenges surrounding increasing climatic events (Reid 2015)

Finally, since achieving independence in the 1980s, Vanuatu's political, governance and economic frameworks have been fragile. Future institutional challenges may lead to political pressures and civil conflicts. The direct and indirect impacts of climate change compound this instability, illustrating Vanuatu's ongoing vulnerability (Buhaug, Gleditsch, and Theisen 2008, 22), including threats to peace and security.

Climate Migration

In Vanuatu, some coastal villages have already been forced to relocate due to the impacts of rising sea levels. For example, in 2005, the coastal community of Lateu on the island of Tegua was moved, as a result of rising sea levels. Vulnerable residents were relocated to the higher grounds of Tegua, after homes were affected by extreme weather, linked to climate change (United Nations 2005). It was noted that the relocation of communities in Tegua was under a project entitled *Capacity Building for the Development of Adaptation in Pacific Island Countries*, which provides an example of a predicted increasing trend of climate migration in the region (ACUNU 2005, 7).

There is a lack of national and international law, policy and planning surrounding climate displacement and migration (Bronen and Pollock 2017). For instance, it is unclear how such relocations should be financed, who should be compensated, where communities should move to, and by when, and who should be responsible for the implications of making such decisions (Schaar 2018, 8). The lack of mechanisms enhances risks and vulnerabilities existing within a nation. For example, resource scarcity in one area may lead to migration to more favourable, resource-abundant areas. Usually, this means internal migration from rural to urban areas, where there may be greater employment opportunities (Gibson and Gurmu 2012). Heightened migratory flows have contributed to the expansion of peri-urban and urban settlements, often comprising housing that has not been designed to withstand extreme climatic events (Keen et al. 2017, 13). The coupling of increased population density in urban centres and inadequate urban planning can have disastrous and long-lasting consequences. Additionally, in Vanuatu, urban settlements are coastal and at risk from more frequent and intensified storm surges, tsunamis, and cyclones due to changes in weather patterns (O'Reilly 2015, 17). This was demonstrated when Cyclone Pam struck Vanuatu in 2015 and significantly damaged the capital city of Port Vila (Wilson 2015).

There may be up to 150 million environmentally displaced people by the end of this century in the Asia region alone (Dunlop and Spratt 2017, 20). Pressures on nature are predicted to escalate in the future, since "[g]lobal warming will drive increasingly severe humanitarian crises, forced migration, political instability and conflict..." (ibid.). This creates unprecedented state and transboundary challenges (United Nations 2009). The loss of territory due to sea level rise threatens sovereignty and territorial integrity, which can create existentialist issues for small island nations, such as Vanuatu. The growing number of uninhabitable islands is resulting in forced migration, which, in turn, intensifies competition for natural resources, and undermines the capacity of state institutions to maintain security. Whether internal (within Vanuatu) or cross-boundary (international), citizens of Vanuatu are increasingly at risk of becoming Environmentally Displaced Persons (EDPs) (Kwa 2008). EDPs are exposed to many risks (Vanuaroroa 2018, preamble) as they migrate internally to urban centres. For example, the informal settlements in the peri-urban areas of Port Vila are expanding, compounding a variety of risks when natural hazards occur (ibid., 11). A causal link can be established when EDPs are forced to leave their homelands due to the environmental and social effects of climate change. Clashes can occur with residents in recipient regions, such as conflicts over scarce resources, employment opportunities and cultural differences (Boege 2015, 2).

Responding to Vanuatu's Climate Vulnerabilities

The second part of this chapter discusses responses to climate change within a peace and security framework. The discussion lists legal and economic possibilities, and culminates in emphasising the importance of empowering local communities.

A Legal Approach

Empowering Kastom

The IPCC has acknowledged the Pacific Islands as "especially vulnerable" to the impacts of climate change (Nurse et al. 2014, 1623). This characterisation has been criticised as implying a lack of confidence in the capacity and agency of its people (Buggy and McNamara 2016, 271). In reality, the opposite is more likely true. Indigenous knowledge, culture and spirituality embody the conservation and preservation of the environment at its core. In Vanuatu, *kastom* is a hybrid of customary laws and traditional ecological knowledge (Forsyth 2004, 429). The notion of *kastom* can be expanded to describe a 'way of life' that is culturally distinctive (Jolly 1992, 341). In recent times, non-Western views have been increasingly heard in climate change discourse. However, these voices tend to be heard in siloed forums rather than at mainstream climate negotiations.

Indigenous ontologies prioritise "relations" over "entities" and embrace humans not only as individuals, but also as defined by: their place within a community, relationship with nature and the spiritual realm. For instance, in Melanesia, personhood is generally understood to be "relational and contextual", rather than individual (Boege 2018, 12). As a result, communities are motivated to address climate change adaptation methods in alignment with cultural and spiritual norms, in order to protect cosmological, as well as environmental assets.

Ni-Vanuatu people understand the importance of coupled human–nature systems to sustain their livelihoods. In Vanuatu, *kastom* not only exists in the abstract, but is cemented in state, regional and local doctrine. Article 95(3) of Vanuatu's Constitution states, "customary law shall continue to have effect as part of the law of the Republic of Vanuatu" (Republic of Vanuatu 1980, art 95(3)). Article 47(1) also declares that the judiciary must determine the "matter according to substantial justice and whenever possible in conformity with custom" (ibid., art 47(1)). As such, Vanuatu operates in a paradigm of legal pluralism, involving both traditional customary law and State law (Jolly 1992, 341). Pluralism has contributed to some of the legal, policy and management strengths and challenges presented by climate change (Davies 2015, 44). For example, in rural areas customary law is often enforced, while in urban areas, State law is more influential.

The Ni-Vanuatu people have long-established adaptation skills which have helped them to prepare for, and recover from, natural disasters (DFA 2017). They have also been provided with contemporary adaptation strategies and skills. For instance, when Cyclone Pam destroyed the majority of the islands' crops, residents were offered assistance with adaptive farming practices, which included: planting climate-resilient species, and community education building a stronger understanding of the effects of climate on traditional agricultural practices (Reid 2015).

A combination of *kastom*, contemporary management principles and resilient ecosystem strategies creates a system of adaptive co-management that will become increasingly important for Vanuatu, in response to growing climate change impacts. This is because communities on a day-to-day level are more likely to be engaged with, and respond positively to, this hybridised system, along with an accompanying structure of by-laws. Indeed, localised solutions are more culturally and environmentally appropriate and empowering than top-down approaches. An educative culture of understanding and respect is essential for reducing disruptions to peace and security. It also establishes a system of mutual accountability and transparency between various stakeholders working towards a common goal of ecosystem protection.

International Law

Historically, Vanuatu advocated for higher levels of global accountability, responsibility and compensation for the South Pacific region. "This includes compensation for damage incurred due to green house gas emissions by developed countries." However, *The United Nations Framework Convention on Climate Change* is limited by the absence of an enforcement mechanism, its political vulnerability, and focus on greenhouse gas (GHG) emissions mitigation. As a result, the UNSC may be arguably the most appropriate body to address the risk of conflict from climate change impacts from the perspective of international law (Schaar 2018, 10). Reframing the impacts of climate change as a potential threat to a nation's peace and security will enable the issue to fall within the ambit of the UNSC and the International Court of Justice (Davies and Riddell 2017).

There are examples of the UNSC considering climate change as a security risk breaching the 'no-harm' international law principle (ibid.). For instance, in the 2017 Resolution on the Lake Chad region conflict, the UNSC acknowledged that it was necessary to recognise "the adverse effects of climate change and ecological changes, among other factors, on the stability of the region" (United Nations 2017a). These included "water scarcity, drought, desertification, land degradation, and food insecurity" (ibid.), with the UNSC highlighting the need for the UN and state governments to implement appropriate risk assessments and risk management strategies. Accordingly, COP24 president Michael Kurtyka noted "local events have a butterfly effect, they impact people's livelihoods, security, ability to provide, to produce, to function. And all through all this they create inflammatory ground on which a conflict can breed" (Mathiesen and Sauer 2019).

Similarly, recent discourse demonstrates an emerging institutional practice by the UNSC towards increasing consideration of non-conflict events as potential threats to international peace and security. This direction could see climate change fall within the UNSC mandate. In a 2005 debate, the UNSC recognised food insecurity as a potential threat to international peace and security (UNSC 2005, 7). Then, in 2014, the spread of the Ebola virus was thought to be a potential trigger for conflict (UNSC 2014, preamble), in what represented the greatest expansion in the scope of the UNSC definition of a threat to peace and security (Nasu 2012, 95). If food insecurity and the spread of disease legitimately constitute a security threat, then it is conceivable that climate change could also fall within this ambit. In fact, as the UNSC progressively mainstreams human security, the potential impacts of climate change have the capacity to hold even greater weight in international environmental law. This added legal force will help to generate more focus on mitigation and adaptation efforts to assist vulnerable nations, such as Vanuatu (Davies and Riddell 2017).

Should the UNSC declare climate change a legitimate threat to international security, there are various measures it may invoke. Article 41 and 42 of the *UN Charter* may allow military or non-military interventions in peace-keeping and humanitarian aid capacities (United Nations 1945). Further, Article 34 authorises the UNSC to "investigate any dispute any situation which might lead to international friction or give rise to a dispute" (ibid.) while Article 35 allows for any UN member to request an investigation (ibid.). Utilising the powers of the UNSC through the *UN Charter* may allow Vanuatu to pursue formal recognition of climate change as a genuine and serious threat to security in the South Pacific and could activate more urgent measures, including availability of resources (Davies and Riddell 2017).

Climate Finance and the Law

Agriculture, fishing and tourism constitute the main components of Vanuatu's economy. These industries are highly dependent on long-term, predictable climatic patterns. Thus, the nation's economic growth and stability are threatened through the less predictable climate as a consequence of climate change. Additionally, building and maintaining infrastructure that can withstand the predicted increased severity and frequency of extreme climatic events is critical for Vanuatu's future to enhance economic growth and reduce poverty in the region. However, this can only be achieved through the financial support of developed countries.

Article 9 of the *Paris Agreement* stipulates developed country parties shall (voluntarily) provide financial resources to assist developing country parties, with respect to both mitigation and adaptation programmes (United Nations 2015, Add.1, art. 9). Further, Article 8 recognises the concept of climate justice by establishing "loss and damage" as an independent pillar of the international climate regime (ibid., art8). Loss and damage acknowledges the inevitable suffering climate change has imposed, and will continue to impose, on vulnerable states, and provides a legal obligation for long-term action. Loss and damage focuses on facilitating strategies, including early warning systems, risk management strategies, insurance facilities, non-economic loss calculation mechanisms and resource-sharing hubs.

However, the concept has been described as an "ambitious compromise" obtained by developed states hoping to implement soft solutions, rather than commit to financial and legal liability schemes (ibid.). The exclusion of liability means the pillar of loss and damage cannot help to overcome the issue of establishing legal causality in climate change actions. Nonetheless, while this exclusion limits the mechanism's scope, it does not displace existing international laws, such as human rights law, world heritage law, the law of the sea, or general international law involving state responsibility. While the *Paris Agreement* provides a solid foundation for action, realising its goals in practice requires enhanced support for vulnerable nations, such as Vanuatu, while simultaneously reducing GHG emissions (United Nations 2016).

As South Pacific neighbours, Australia and New Zealand have a responsibility to support Vanuatu. Greater collaboration and collective action among all actors are necessary to improve access to climate finance. Australia dedicated $3 billion AUD towards infrastructure in the South Pacific, with then Prime Minister Scott Morrison noting "Australia has an abiding interest in a south-west Pacific that is secure strategically, stable economically and sovereign politically" (Wroe 2018). While such financial support is promising, greater pressure could be placed on neighbouring developed states. Applying the international environmental laws principle of *Common but Differentiated Responsibilities and Respective Capabilities* acknowledges the disproportionate contribution of adverse human-induced climatic harm that can be attributed to developed states. It helps to distribute responsibility, and places the onus on developed states to provide remedies for vulnerable nations, such as Vanuatu (Khare 2016, 99). A further principle of international law that is increasingly underpinning climate litigation is *Intra- and inter-generational equity and justice*, although it is yet to be codified. The application of this principle is important to the legacy of climate change on future generations of Ni-Vanuatu citizens and their rights to a peaceful and secure future.

Localised Approaches

Policy

The value of *kastom*, international environmental law and climate finance as a means of addressing the impacts of climate change in Vanuatu would be diminished in the absence of adequate policy, education and communication tools (Republic of Vanuatu 2017, foreword). In the light of this, a nationwide '*Climate Change and Disaster Risk Reduction Policy*' was developed to enhance the resilience and awareness of the Ni-Vanuatu people. Based on six founding principles (accountability, sustainability, equity, community focus, collaboration and innovation), the policy emphasises transparency among stakeholders (ibid., 2). The policy excludes conflict prevention mechanisms, relying on the founding principles to ensure that peace is maintained in climate change actions.

Education

Education programmes taught at schools and in communities are helpful in providing strategies to prevent potential hostility (Kotite 2012) particularly when based on 'lessons learnt' from previous disasters and disaster risk reduction schemes (Republic of Vanuatu 2017, 15). Efforts are best invested in establishing and maintaining open and accessible channels of communication, where educators can disseminate information and build networks to link shared resources widely (ibid., 14). Tensions in Vanuatu, which may lead to conflict, can be alleviated between communities and government agencies, through open and accessible modes of communication, embedded trust and good governance.

Local Management and Leadership

The Pacific Islands receive more aid per capita than any other region in the world, some of which often fails to reach local levels (Pawar 2009, 77). Academics have long recognised top-down approaches to aid policy that are driven by external agencies and experts, rather than local needs, as "wholeheartedly ineffective" (Buggy and McNamara 2016, 271). Issues of mismanagement, maintenance and mistrust have been behind historic aid project failures (ibid., 270), which may clash with long-enshrined traditions, customs and power relations. While it is important to acknowledge the need for international and national frameworks, communities across the Pacific Islands have demonstrated an emerging resolve to prioritise localised measures to support climate change adaptation. This 'grass roots' approach will assist the early mitigation of climate related conflicts.

Social marketing research affirms that people are more likely to take action on an issue when they feel a strong sense of affiliation with the person, or institution, making the request (CRED 2014). According to the Centre for Research on Environmental Decisions, "local leaders may be more likely to set a norm for climate action than calls for action from people outside the community" (ibid., 10). Learning from community leaders enables the inclusion, and bridging, of social and cultural barriers (Amel et al. 2017, 277). Utilising trust relationships within established community structures helps to break down existing psychological-cognitive barriers to addressing the impacts of climate change (Moser 2009). These lessons support the importance of empowering local Vanuatu leaders, such as Chiefs, in addressing the impacts of climate change, together with the maintenance of peace.

Climate Change as a Community Connector

Previous examples, such as the Civil War in Syria, demonstrate how the impacts of climate change can bring people together in 'negative' and 'destructive' frameworks of conflict. There are opportunities to reverse this scenario by utilising the 'negative' impacts of climate change in a constructive way, by driving 'positive' community connectors and connections. Community cohesion and support, on local, national and international levels, can result when experiences and stories are shared by affected communities. Some of the most pertinent voices are those being heard from young people in terms of the intergenerational injustices presented by the changing climate. For example, when central and eastern Australia was declared a drought zone in 2018, a fundraising campaign started by a 10-year-old boy raised more than $1 million in less than five weeks (Duncan and Fitzpatrick 2018). The Muslim Charity Foundation delivered 33 tonnes of hay to Brisbane farmers (Lawira 2018) and Coles and Woolworths supermarket chains matched dollar-for-dollar customer donations to drought-relief schemes (AAP 2018). Activating stronger community connections and support for Vanuatu, particularly that of developed states, will require the increased dissemination of climate change narratives.

Adaptation Coalitions

Adaptation Coalitions form a key part of the approach to climate change adaptation, involving the integration of science and local knowledge (Smith and Vivekananda 2007, 29). Adaptation Coalitions are "community groups that come together as an internal coalition and form alliances with outside groups in order to achieve common desired futures around climate change vulnerability and impacts" (Ashwill, Flora, and Flora 2011, vii). The foundation of the *Adaptation Coalition Framework* (ACF) relies on strengthening internal organisations to take action, by building social capital, then linking these empowered groups to goal-aligned external partners. For example, more than 25 communities across five Latin American states have trialled Adaptation Coalitions with significant success marked by the mobilisation of local personnel, access to appropriate financial and material resources, and agreements made between communities and external institutions (ibid.). Increasing aid flows does not translate into more effective outcomes (Buggy and McNamara 2016, 271); investment in social capital, along with financial capital, holds greater promise for the Pacific Islands, including Vanuatu.

Conclusion

This policy brief has outlined why Vanuatu is highly vulnerable to the impacts of climate change and how these vulnerabilities threaten the nation's peace and security. The 'perfect storm' is brewing as the nation's population grows and its exposure to climate risks escalates, as the planet continues to warm. Of particular concern is the future of rural citizens who constitute the majority of the nation's population. This brief provides the following recommendations that span local, national and international jurisdictions and have been designed to address some of these mounting concerns. Embedded in each recommendation is the aim to maintain peace and security by minimising the likelihood of conflict.

Policy Recommendations

1 *International: UNSC Recognising Climate Change as a Security Threat*
   International frameworks need to respond adequately to the threats posed by the impacts of climate change. The UNSC's recognition of climate change as a threat to the security of vulnerable nations could prompt necessary mitigation and adaptation measures to prevent climate change induced conflict.
   Recommendation: Vanuatu may resolve to bring their case, pertaining to the damages and losses the nation has incurred through the impacts of climate change, before the UNSC. Their case could be based on the threats imposed on the nation's current and future viability, including threats to its peace and security.

2 *National: Policy on Resettlement and Internal Displacement*

At a national level, a policy on the resettlement and internal displacement is required. Planned relocation is an option of last resort, when communities have no alternative other than to leave their home localities which are no longer viable. The policy will ensure that lessons learnt from previous relocation experiences globally, and in the Pacific, are considered, so that movement takes place with dignity, appropriate safeguards and human rights protections (Vanuaroroa 2018, 7). To achieve this requires extensive preplanning for the relocated and recipient communities. Key to the success of such a policy and its implementation is the identification of resources to facilitate relocation, together with the evaluation of displacement risks, including those likely to lead to conflict (ibid., 8).

Recommendation: Develop a national policy on the resettlement and internal displacement for the climate displaced citizens of Vanuatu.

3 *Local: Community-Led Approach*

The potential for conflict is driven by compelling climate drivers. For example: food insecurity due to changing climatic patterns, and the inundation of land due to sea level rise and erosion. These vulnerabilities vary in type and levels of influence. They may lead to conflict including the forced environmental displacement of peoples. The key is to be able to identify and mitigate these vulnerabilities in the pre-phase to conflict, in the interests of maintaining peace and security. Vanuatu communities in rural areas are skilled in adapting to changing environmental conditions and living within the bounds of nature. Cultural issues at the grass roots level can both bring communities together and/or cause conflict between groups (Karthik 2014, 7). Therefore, it would be advantageous to incorporate *kastom*, which encapsulates traditional customary law and knowledge, as a platform, when developing contemporary climate adaptation measures in Vanuatu (ibid.). Empowering local leadership, such as through the Chiefs and education programmes, can also become powerful tools in the mitigation of conflict (MJCS 2016, 6).

Recommendation: Identify conflict and climate vulnerabilities, and their levels of influence at a local level. Based on this evaluation, develop risk reduction approaches to ensure the maintenance of peace. The incorporation of traditional culture (*kastom*), knowledge and local leadership (Chiefs) will ensure that initiatives are appropriate and engage local communities.

4 *Local to International: Climate Change as a Community Connector*

There are opportunities to reframe the 'negative' impacts of climate change in a constructive way, as a 'positive' community connector. This approach offers the benefits of activating stronger global community connections and support for Vanuatu, particularly the support of developed states.

Recommendation: Increase the dissemination of Vanuatu's local climate change narratives. This will build international awareness of the local plight of the nation by utilising climate change as a community connector.

References

ACUNU. 2005. "Worldwide Emerging Environmental Issues Affecting the US Military". Fort Belvoir: AEPI.

Amel, Elise, Christine Manning, Brian Scott, and Susan Koger. 2017. "Beyond the Roots of Human Inaction: Fostering Collective Effort toward Ecosystem Conservation." *Science* 356 (6335): 275–279.

Ashwill, Maximillian, Cornelia Flora, and Jan Flora. 2011. "Building Community Resistance to Climate Change: Testing the Adaptation Coalition Framework in Latin America." Report, the International Bank for Reconstruction and Development. Washington: The World Bank.

Australian Associated Press (AAP). 2018. "Coles, Woolworths: Supermarkets Fundraising for Drought Relief." *News.com*, August 9, 2018. https://www.news.com.au/finance/business/retail/coles-to-match-dollarfordollar-customer-donations-to-drought-relief/news-story/88d7841cac724b0a00a67a7c44bebf09

Blondel, Alice. 2012. "Climate Change Fuelling Resource-Based Conflicts in the Asia-Pacific." Asia Pacific Human Development Report: Background Paper Series No 12, New York: United Nations Development Programme.

Boege, Volker. 2015. "Climate Change, Migration (Governance) and Conflict in the South Pacific." CLISEC Working Paper 29. Hamburg: University of Hamburg Research Group Climate Change and Security.

Boege, Volker. 2018. "Climate Change and Conflict in Oceania: Challenges, Responses and Suggestions for a Policy-Relevant Research Agenda." Toda Peace Institute Policy Brief No.17. Tokyo: Toda Peace Institute.

Bronen, Robin, and Denise Pollock. 2017. "Climate Change, Displacement and Community Relocation: Lessons from Alaska." Oslo: Norwegian Refugee Council.

Brzoska, Michael. 2018. "Weather Extremes, Disasters, and Collective Violence: Conditions, Mechanisms, and Disaster-Related Policies in Recent Research." *Current Climate Change Reports* 4 (4): 320–329.

Buggy, Lisa, and Karen Elizabeth McNamara. 2016. "The Need to Reinterpret "Community" for Climate Change Adaptation: A Case Study of Pele Island, Vanuatu." *Climate and Development* 8 (3): 270–280.

Buhaug, Halvard. 2018. "Global Security Challenges of Climate Change." Toda Peace Institute Policy Brief No.18. Tokyo: Toda Peace Institute.

Buhaug, Halvard, Nils Gleditsch, and Ole Theisen. 2008, "Implications of Climate Change for Armed Conflict." Working Paper. Washington DC: World Bank Group.

Centre for Research on Environmental Decisions (CRED). 2014. *Connecting on Climate: A Guide to Effective Climate Change Communication*. New York: Earth Institute, Colombia University.

Davies, Kirsten. 2015. "Ancient and New Legal Landscapes: Customary Law and Climate Change, a Vanuatu Case Study." *Asia Pacific Journal of Environmental Law* 18: 43–67.

Davies, Kirsten, and Thomas Riddell. 2017. "The Warming War: How Climate Change is Creating Threats to International Peace and Security." *Georgetown Environmental Law Review* 30 (1): 47–74.

Department of Defense (DoD). 2014. "2014 Climate Change Adaptation Roadmap". Working Paper. Washington DC: Department of Defense United States of America.

Department of Foreign Affairs (DFA). 2017. "Development Assistance in Vanuatu: Building Resilient Infrastructure and an Environment for Economic Opportunity in

Vanuatu."https://dfat.gov.au/geo/vanuatu/development-assistance/Pages/building-resilient-infrastructure-vanuatu.aspx

Duncan, Sam, and Stephen Fitzpatrick. 2018. "Fiver for a Farmer Campaign Raises $1m for Drought Relief." *The Australian*, September 8, 2018. https://www.theaustralian.co.au/national-affairs/fiver-for-a-farmer-campaign-raises-1m-for-drought-relief/news-story/be1bae02225aa5cfed8e98b92c0f7e04/

Dunlop, Ian, and David Spratt. 2017. "Disaster Alley: Climate Change, Conflict and Risk." Melbourne: Breakthrough. Available at https://www.breakthroughonline.org.au/disasteralley

Espiritu Santo Tourism Association (ESTA). 2018. "Welcome to Espiritu Santo." http://www.espiritusantotourism.com

Forsyth, Miranda. 2004. "Beyond Case Law: Kastom and Courts in Vanuatu." *VUW Law Review* 35 (2): 427–446, http://www5.austlii.edu.au/nz/journals/VUWLawRw/2004/15.html#fnB47

Garschagen, Matthias, Michel Hagenlocher, Martina Comes, Mirjam Dubbert, Robert Sabelfeld, Yew Jin, Lee, Ludwig Grunewald et al. 2016. *World Risk Report 2016*. Berlin: Bündnis Entwicklung Hilft and UNU-EHS.

Gibson, Mhairi, and Eshetu Gurmu. 2012. "Rural to Urban Migration is an Unforeseen Impact of Development Intervention in Ethiopia." *PLoS ONE Journal* 7 (11), https://doi.org/10.1371/journal.pone.0048708.

Gubb, Matthew. 1994. "Vanuatu's 1980 Santo Rebellion: International Responses to a Microstate Security Crisis." Canberra Papers on Strategy and Defence No 107. Canberra: Australia National University.

Intergovernmental Panel on Climate Change (IPCC). 2018. "Global Warming of 1.5° C (Summary for Policymakers)' (Special Report, Intergovernmental Panel on Climate Change, 8 October 2018) available from https://www.ipcc.ch/sr15/

Johnson, Johanna, David Welch, and Adam Fraser, eds. 2016. "Climate Change Impacts in North Efate, Vanuatu" (Working Paper) New Caledonia: Communauté du Pacifique.

Jolly, Margaret. 1992. "Custom and the Way of the Land: Past and Present in Vanuatu and Fiji." *Oceania* 62 (4), 330–354.

Karthik, K. 2020. "Consequence of Cross Cultural Misunderstanding – A Shipboard Perspective." *Indian Journal of Science and Technology*, 7 (S7): 6–9. https://doi.org/10.17485/ijst/2014/v7sp7.4

Keen, Meg, Julien Barbara, Jessica Carpenter, Daniel Evans, and Joseph Foukona. 2017. "Urban Development in Honiara Harnessing Opportunities, Embracing Change." Research Report. Canberra: Australia National University.

Kelley, Colin P., Shahrzad Mohtadi, Mark A. Cane, Richard Seager, and Yochanan Kushnir. 2015. "Climate Change in the Fertile Crescent and Implications of the Recent Syrian Drought." *Proceedings of the National Academy of Sciences* 112 (11): 3241–3246.

Khare, Rishika. 2016. "The Principle of 'Common but Differentiated Responsibilities' and the Challenges Posed by it in the Context of International Climate Governance." *International Journal of Law and Legal Jurisprudence Studies* 3 (2): 98–113.

Kotite, Phyllis. 2012. "Education for Conflict Prevention and Peacebuilding: Meeting the Global Challenges of the 21st century." IIEP's Occasional Paper. Paris: United Nations Education, Scientific, and Cultural Organization.

Kwa, Eric. 2008. "Climate Change and Indigenous People in the South Pacific." Paper presented at International Union for Conservation of Nature Academy of

Environmental Law Conference on Climate Law in Developing Countries post-2012: North and South Perspectives, Ottawa, 26–28 September.

Lawira, Laurie. 2018. "Muslim Charity Donates Three Truckloads of Hay to Drought-Stricken Farmers." *SBS News*, September 17, 2018. https://www.sbs.com.au/news/muslim-charity-donates-three-truckloads-of-hay-todrought-stricken-farmers

MacQueen, Norman. 1980. "Beyond Tok Win: The Papua New Guinea Intervention in Vanuatu, 1980". *Pacific Affairs* 61: 235–252.

Mathiesen, Karl, and Natalie Sauer. 2019. "UN Security Council Members Mount New Push to Address Climate Threat." *Climate Home News*, Published 25 January 2019. https://www.climatechangenews.com/2019/01/25/un-security-council-members-mount-new-push-address-climate-threat/

Ministry of Justice and Community Services (MJCS). 2016. "Conflict Management and Access to Justice in Rural Vanuatu." Working Paper. Canberra: Department of Foreign Affairs and Trade.

Moser, Susanne. 2009. "Communicating Climate Change - Motivating Civic Action: Renewing, Activating and Building Democracies?" MIT Press Scholarship Online. https://doi.org/10.7551/mitpress/9780262012997.003.0014

Mudaliar, M.M., C. Bergin, and K. MacLeod. 2015. "Drinking Water Safety Planning: A Practical Guide for Pacific Island Countries." Suva: World Health Organization and Pacific Islands Applied Geoscience Commission.

Nasu, Hitoshi. 2012. "The Place of Human Security in Collective Security." *Conflict Security* 18 (1): 95–129.

Nurse, Leonard, Roger McLean, John Agard, Lino Pascall Briguglio, Virginie Duvat-Magnan, Netatua Pelesikoti, Emma Tompkins, et al. IPCC. 2014. "Small Islands". In *Climate change 2014. Impacts, Adaptation, and Vulnerability. Part B: Regional Aspects*. Contribution of Working Group II Contribution to the Fifth Assessment Report of the Intergovernmental Panel on Climate Change, edited by V.R. Barros, C.B. Field, D.J. Dokken, M.D. Mastrandrea, K.J. Mach, T.E. Bilir, M. Chatterjee, K.L. Ebi, Y.O. Estrada, R.C. Genova, B. Girma, E.S. Kissel, A.N. Levy, S. Mac-Cracken, P.R. Mastrandrea, and L.L. White, 1613–1654. Cambridge: Cambridge University Press.

O'Reilly, Dominic, ed. 2015. "Urbanization and Climate Change in Small Island Developing States." Report for the United Nations, Human Settlements Programme. Nairobi: UN-Habitat.

Pawar, Manohar. 2009. *Community Development in Asia and the Pacific*. London: Routledge.

Reid, Kathryn, 2015. "2015 Cyclone Pam: Facts, FAQS, and How to Help." Accessed 3/7/2018. https://www.worldvision.org/disaster-relief-news-stories/cyclone-pam-facts

Republic of Vanuatu. 1980. Constitution of the Republic of Vanuatu.

Republic of Vanuatu. 2017. "Vanuatu National Environment Policy and Implementation 2016–2030." Working Paper. Port Vila: Government of the Republic of Vanuatu.

Reti, Muliagatele Joe. 2007. "An Assessment of the Impact of Climate Change on Agriculture and Food Security: A Case study in Vanuatu." Working Paper. Apia: Pacific Environment Consultants Ltd.

Schaar, Johan. 2018. "The Relationship between Climate Change and Violent Conflict." Working paper. Stockholm: Sida.

Scheffran, Jürgen, Michael Brzoska, Jasmin S.A. Link, Peter Michael Link, and Janpeter Schilling. 2012. "Disentangling the Climate-Conflict Nexus: Empirical and

Theoretical Assessment of Vulnerabilities and Pathways." *Review of European Studies* 4 (5):1–13.

Secretariat of the Pacific Regional Environment Programme (SPREP). 2017. "Vanuatu National Environment Policy and Implementation 2016–2030." Working Paper, Government of the Republic of Vanuatu. Apia: SPREP.

Smith, Dan, and Janani Vivekananda. 2007. *A Climate of Conflict: The Links between Climate Change, Peace and War.* London: International Alert.

Sternberg, Troy. 2012. "Chinese Drought, Bread and the Arab Spring." *Applied Geography* 34 (May): 519–524.

UNICEF. 2017. "Thirsting for a Future: Water and Children in a Changing Climate". Report. New York: UNICEF.

United Nations. 1945. *Charter of the United Nations.* https://www.un.org/en/sections/un-charter/un-charter-full-text/

United Nations. 2005. "UN Environmental Body Hails Relocation of Islanders Threatened by Climate Change." *UN News*, 6 December 2005. https://news.un.org/en/story/2005/12/162492-un-environmental-body-hails-relocation-islanders-threatened-climate-change

United Nations. 2009. UN General Assembly, *Climate Change and Its Possible Security Implications:* UN GAOR, 64th Sess., Provisional Agenda Item 144, UN Doc A/64/350 (11 September 2009).

United Nations. 2010. Sustainable Development, *Vanuatu National Assessment Report: 5 Year Review of the Mauritius Strategy for Further Implementation of the Barbados Programme of Action for Sustainable Development.* https://sustainabledevelopment.un.org/content/documents/1380Vanuatu-MSI-NAR2010.pdf

United Nations. 2015. Conference of the Parties, United Nations Framework Convention on Climate Change, *Report of the Conference of the Parties on Its 21st Session, Held in Paris from 30 November to 11 December 2015 – Addendum – Part 1* UN Doc FCCC/CP/2015/10/.

United Nations. 2016. United Nations: Climate Change. https://unfccc.int/topics/climate-finance/the-big-picture/climate-finance-in-the-negotiations

United Nations. 2017a. "Department of Public Information." *Security Council Strongly Condemns Terrorists Attacks, Other Violations in Lake Chad Basin Region, Unanimously Adopting Resolution 2349.* SC/12773, 31 March 2017. https://www.un.org/press/en/2017/sc12773.doc.htm

United Nations. 2017b. "Department of Public Information." *Security Council Holds First-Ever Debate on Impact of Climate Change on Peace.* SC/9000, April 17, 2007. https://www.un.org/press/en/2007/sc9000.doc.htm

United Nations. 2017c. *UN Climate Change Conference: Vanuatu* (2017) United Nations COP 23 Fiji https://cop23.com.fj/vanuatu/

United Nations. 2018. Deputy Secretary General. "Impacts of Climate Change Go Well Beyond 'the Strictly Environments'" [Environmental Issues and Sustainable Development], DSG/SM/1195-SC/13418-ENV/DEV/1861, United Nations, 11 July 2018, www.un.org/press/en/2018/dsgsm1195.doc.htm

UNSC. 2005. 60th Sess., 5220th meeting held June 30, 2005. https://undocs.org/en/S/PV.5220

UNSC resolution 2177, S/RES/2177 (18 September 2014), available from http://unscr.com/en/resolutions/doc/2177

Vanuaroroa, Ham Lini. 2018. *National Policy on Climate Change and Disaster-Induced Displacement.* Vanuatu: NDMO.

Wilson, Catherine. 2015. "Cyclone Pam Worsens Hardship in Port Vila's Urban Settlements." *Inter Press Service*, April 13, 2015. http://www.ipsnews.net/2015/04/cyclone-pam-worsens-hardship-in-port-vilas-urban-settlements

Wong, P.P., I.J. Losada, J.-P. Gattuso, J. Hinkel, A. Khattabi, K.L. McInnes, Y. Saito, and A. Sallenger, 2014. "Coastal systems and low-lying areas." In *Climate Change 2014: Impacts, Adaptation, and Vulnerability. Part A: Global and Sectoral Aspects. Contribution of Working Group II to the Fifth Assessment Report of the Intergovernmental Panel on Climate Change*, edited by Field, C.B., V.R. Barros, D.J. Dokken, K.J. Mach, M.D. Mastrandrea, T.E. Bilir, M. Chatterjee, K.L. Ebi, Y.O. Estrada, R.C. Genova, B. Girma, E.S. Kissel, A.N. Levy, S. MacCracken, P.R. Mastrandrea, and L.L. White, 361–409. Cambridge and New York: Cambridge University Press.

Wroe, David. 2018. "Scott Morrison Splashes Cash in the Pacific as China Fears Loom." *The Sydney Morning Herald*, November 8, 2018. https://www.smh.com.au/politics/federal/scott-morrison-splashes-cash-in-the-pacific-as-china-fears-loom-20181107-p50emv.html

9 Conclusions

Volker Boege

What the contributions in this book have clearly demonstrated is the necessity for a holistic and inclusive approach to the problem of climate change, conflict, security and peace. The "climate-fragility risks" that are talked about a lot in the academic community working on the problem (see, for example, Rüttinger et al. 2015; Vivekananda et al. 2017) cannot be confined to the fragility of states. Of course, it is significant both from an academic and a political perspective to explore and address the link between fragile states and climate change (in both directions: climate change enhancing state fragility, and state fragility hampering adequate responses to climate change), but it is not enough—the challenge is much more complex, it reaches wider and deeper. The case study chapters in this book, in addition to political fragility, have also addressed climate change-induced societal, emotional and spiritual fragility. In Chapter 5, John Campbell talks about "ontological security" as a more inclusive approach that also comprehends these societal, emotional and spiritual aspects. It will be worthwhile indeed to take up the ontological security discourse from the perspective of sociology, social psychology and peace and conflict studies and explore its potential for coming to terms with the fundamental challenges posed to peace and security by climate change.[1]

The Pacific Islands Forum's declaration on security in the region, the Boe Declaration, of September 2018, which posits that climate change is "the single greatest threat to the livelihoods, security and wellbeing of the peoples of the Pacific" (Boe Declaration 2018) points in this direction when it talks about "human security" and an "expanded concept of security" (ibid.), but it still falls short of coping with the actual challenges. In the light of the unprecedented dangers of climate change, a focus on human security as the security of human beings and human societies in isolation, as separated from the non-human 'rest' of the world (or of creation, if you like) will not suffice; rather, this anthropocentric way of seeing the world and thinking about it, this reification of a divide between human society on the one hand and 'nature' on the other is a cause of the climate change-induced emergency in which mankind finds itself (having created that emergency).

DOI: 10.4324/9781003001744-12

The Toda Pacific Declaration on Climate Change, Conflict and Peace of July 2019 is an attempt at a more far-reaching alternative, holistic and inclusive approach, drawing attention to non-Western, non-anthropocentric, relational concepts and to Indigenous knowledge and indigenous Pacific ways of climate change adaptation, of conflict transformation and of peacebuilding. The Declaration advocates to overcome "human-centred approaches, which separate people from nature, nurturing the concept of relationality which will deepen connections between people and other living beings and the material and immaterial worlds", and it recommends a focus on "dimensions of the climate change–conflict nexus which so far have been widely ignored or underestimated, such as cultural and spiritual aspects, gender, traditional customary law and knowledge" (Toda Declaration 2019).

Such an understanding of the problematique of climate change, conflict, peace and security is worlds apart from certain positions in the climate change/security discourse which Matt McDonald discusses in his contribution to this book. He succinctly traces the debate about the securitisation of climate change and explores the pros and cons of such securitisation: it can be seen, he argues, as elevating climate change to the 'high politics' of security and thus securing more political attention and more funding for addressing it, or it can be seen as opening the door for the militarisation of responses to climate change and giving the defence and military establishments new opportunities to demand additional resources and to strengthen their legitimacy. As Matt McDonald points out, 'security' comes in various shades when linked to climate change: national security, international security, human security, ecological security. In other words, this is a highly contested field.

On the one pole, linking climate change and security can lead to the militarisation of the topic and to military responses, preparing one's military forces for interventions in 'fragile situations' in regions of the Global South ravaged by climate change and its conflictive social and political effects, or it can lead to the framing of climate-changed induced migration as a 'security threat', with the response of the fortification of one's own borders in protection against the 'waves' of climate refugees. Such narratives present victims of climate change as threats to national security and legitimise responses to climate change based on force; in this context, the military is presented or can present itself once more as major provider of—national—security, now with regard to climate change as a security threat. A recent example of such an approach is the 'Worldwide Threat Assessment' of the CIA, FBI and other US agencies of January 2019 which states that "climate change is an urgent and growing threat to our national security" (quoted from Nuccitelli 2019). This is in line with the Pentagon's perspective on climate change as a security issue (Klare 2019).

On the other pole of the debate there are approaches, like the Toda Pacific Declaration, that try to fill the concept of security relationally, even transcending the anthropocentric understanding of peace and security which is still implicit in 'human security' and the divide between (human) society and (non-human) nature. Concepts like 'ecological security', 'ontological security' or 'community security' point in this direction—with 'community' in the most far-reaching concepts understood as not only comprising living human beings, but also unborn generations, spirits (of the ancestors), other living beings and elements of the material and immaterial world, all of which do not exist as distinct entities, but only relationally (I will come back to this later).

As Matt McDonald in his contribution makes clear, the debate is still wide open, with a vast variety of positions and approaches between the two poles outlined above. It is worthwhile to engage in this contestation about the meaning of security to make Pacific voices and peace research voices heard. This means, for example, to try to exert influence on the efforts to address climate change and security in the context of the United Nations. In the UN context, the topic of climate change, security and peace has gained considerably in prominence over the last years. From 2007 onwards, for example, the UN Security Council has held several open debates on climate change as a security threat, and a modest Climate Security Mechanism was established in the UN system in October 2018, hosted by the UN Department for Peace-building and Political Affairs, with input from the United Nations Development Programme and the United Nations Environment Programme. Its task is to work on climate-related security risk assessments, based on a security risk assessment framework which is to capture risks in different world regions. The Pacific Island countries have been particularly active in these debates and activities; most of them are members of the UN 'Group of Friends' on climate security which was established in August 2018, jointly led by Germany and the Pacific Island country of Nauru, and which is constantly growing in membership. Germany has announced that during its presidency of the UNSC in 2020 it will have a special focus on climate change and security. The PICs promote even more ambitious plans—for example, they have requested that the UN Secretary General appoint a Special Adviser on Climate Change and Security, and the UN Security Council appoint a Special Rapporteur to elaborate a regular review of security threats posed by climate change (see, for example, Pacific Islands Forum 2019, 15).

The contributions to this book demonstrate that Pacific views and experiences have an important contribution to make to the international debate, in the UN context and beyond. And they show that, based on Pacific experiences, attention has to be drawn to issues and problems which in the mainstream debate so far have not mattered much.

To begin with, the linkages between climate change and violent conflict/violence need to be explored at scales and from viewpoints which so far have been neglected or underestimated. The case studies in this book show that it is necessary to look much more and much closer at the everyday and at the local level when exploring climate-related security threats, violence and violent conflict—well beyond large-scale violent conflicts, international or civil wars and the like. It is here in the local everyday context that people are severely affected. Matt McDonald in his contribution reminded us that climate-related livelihood loss, disease, insufficient food and water (that is, the effects of structural violence) are more immediate, more direct and more pressing threats to people's everyday security than the danger of armed conflict. And, as I pointed out in my chapter, if there is violence, it is often small-scale or low-level, for example, domestic violence, violence between and within communities, violent conflict over access to scarce resources in a local context or between relocated and recipient communities. This confirms Halvard Buhaug's assessment in his chapter of the empirical findings of research on climate change and violent conflict: communal violence or urban riots, not large-scale (civil) war are (at least for the time being) the most frequent forms of climate change-related violent conflict. These forms of everyday and localised violence mostly remain under the radar of mainstream research on climate change and violent conflict; they are certainly not captured by the current dominant quantitative research approach in the study of the climate–conflict nexus.[2] Furthermore, the chapters in this book have shown that increased risk of violence or violent conflict in the climate change—conflict context is dependent on the fragility or stability of the overall societal and political context (see also Rüttinger et al. 2015; Nett and Ruettinger 2016), and they have put climate change into perspective as one among other factors contributing to violent escalation of conflict. One has to take into account other factors and their relevance; as Halvard Buhaug and Matt McDonald explain, a sole focus only on climate change-related security threats can be misleading. Similarly, in my chapter, I make the point that the significance of the factor climate change for conflict and violent escalation of conflict has to be situated in the specific case-dependent complex conflict constellation and its emergence over time. In other words: pathways and context matter.

Second, the chapters of this book have confirmed the significance of the climate change–migration–conflict nexus, but at the same time have cast new light on it. Halvard Buhaug in his chapter identifies forced migration as a prominent pathway between climate change and armed conflict, and he posits that gaining better insights into how climate change interacts with other drivers of human mobility and its knock-on security impacts should be a key priority for future research. John Campbell's chapter on climate change, land and migration, as well as the case studies in this book, have made it perfectly clear that various forms of human mobility—relocation, displacement and migration—and their conflict potential have to be understood in a much more comprehensive way than is usually the case in climate change–migration–conflict studies. Research that merely looks at this issue as a technical, economic and—maybe—political one, and practice that merely addresses it as such, falls short of the challenge and hence risks both missing its conflict-proneness and failing in conflict prevention. Chapters by John Campbell and others stress the significance of the land/identity connection in Pacific Island societies and cultures.

Kate Higgins explicitly makes the point that land should not be understood only as the physical location where people live or as an economic asset, but also in terms of its social, relational, cosmological and spiritual dimensions. Following from such an understanding, it becomes clear that identity for Pacific Islanders is inextricably linked to land, to the place of origin. This connection to land is under threat due to the effects of climate change (sea level rise, coastal erosion, food insecurity etc.), and hence, so too is the land-based identity of people. John Campbell therefore stresses that climate change and climate change-induced migration does not only have serious implications for the material security of affected communities, but also affects their emotional and spiritual wellbeing, their ontological security. This is confirmed by the Boege and Rakova chapter with regard to the resettlement of Carteret Islanders to mainland Bougainville. People are torn between the desire to stay put, to maintain their connection to the land, and the need and insight to plan for resettlement in the increasing number of cases in which ancestral land—low lying islands, coastal areas—become uninhabitable or even inundated. Climate change-induced displacement, migration and relocation pose a major challenge to people's identity which is rooted in their place of birth and connected to the ancestors of that place. John Campbell even goes so far as to say that loss of customary land for community groups equals the loss of an important rationale for their existence, with severe emotional, psychological and spiritual implications. Conflicts arising from this challenge are to a large extent identity conflicts, within communities (for example, gender-based or domestic violence), but also between relocating and recipient communities (see also Brzoska and Fröhlich 2015). The case study chapters provide examples for such conflicts: the Boege and Rakova chapter talks about conflicts between relocated people and recipient communities, the chapter by Kate Higgins points to violence in the informal settlements in the capital city of the Solomon Islands, Honiara; these settlements are sites of climate-induced in-migration that at the same time are particularly vulnerable to climate change impacts and natural disasters.

Accordingly, research that wishes to explore the climate change–migration–conflict nexus will have to ask questions like: what does constitute identity in specific sites and social-cultural contexts?, in which ways is identity challenged due to the effects of climate change?, how do people cope with challenges to identity in the course of (imminent, current or past) climate-change induced displacement, relocation and migration?, and what does this mean for the conflict-proneness of the effects of climate change, the conduct of actual conflicts, and prospects for conflict prevention, conflict transformation and peacebuilding?

A third point worth highlighting is that conflict-sensitive and conflict-preventive climate change governance, policies and strategies have to be grounded in a broad network of actors and institutions. The fixation on governments, state institutions, international and regional organisations is not helpful. Even the inclusion of civil society in the Western understanding of the term is not sufficient. In places like the communities of the Pacific Island countries, actors and institutions which more often than not fail to figure in mainstream research and policies are of utmost importance for the success of climate change adaptation policies and projects. In my contribution I have pointed to the importance of 'unconventional' actors and institutions of climate change governance, such as customary law, traditional Indigenous knowledge, traditional authorities from the customary sphere of societal life (chiefs, elders, priests, male and female holders of traditional Indigenous knowledge, …), or church leaders. I have explained that, in the Pacific, these traditional authorities as well as the churches are to a large extent in charge of the governance of communities and natural resources and for conflict prevention, dispute resolution and peacebuilding, and, accordingly, for conflict-sensitive and conflict-preventive climate change governance and adaptation, are following custom, customary law and traditional Indigenous knowledge. The case studies presented in this book confirm this argument. Kate Higgins in her chapter highlights the importance of customary negotiations, for instance, about relocation or over resource rights, and she gives various examples of the significance of customary leadership for climate change adaptation in the Solomon Islands. The Boege and Rakova chapter also points to the importance of forging relations between relocating and recipient communities in customary ways, and Kirsten Davies recommends the incorporation of 'kastom'—traditional customary law and knowledge—and local traditional leadership into the development of climate adaptation measures in Vanuatu, arguing that this will ensure that initiatives are culturally appropriate and will actually engage local communities. In this context, Kate Higgins is highly critical of establishing new 'climate change' or 'adaptation' institutions and instead recommends working with already existing embedded and legitimate institutions such as churches, chiefs, elders and other forms of local leadership. With regard to the situation in Vanuatu, Kirsten Davies finds that localised solutions are more culturally and environmentally appropriate and empowering than top-down approaches. This very much resonates with Jenny Bryant-Tokalau's plea of "listening to local communities and respecting knowledge that already exists" (Bryant-Tokalau 2018, 3). In short, stressing the need to engage with existing agency, adaptive capacity and community mechanisms, to build on the existing adaptive capacities of communities and to work with local systems of governance and justice, is a common thread that runs through the case study chapters.

At the same time the case study chapters stress the need to combine the efforts of actors and institutions from the state, civil society and the customary sphere—and the difficulties of doing so. Kate Higgins in her chapter sketches the frictions in the relationships between rural communities and highly centralised state institutions in the case of the Solomon Islands when it comes to the implementation of climate change adaptation policies; flowing from that, she recommends shifting the focus to state-community relations instead of just seeing the state and its development partners from the Global North as 'delivering' adaptation solutions to the communities, from the top and from the outside as it is. The Boege and Rakova chapter points to similar difficulties in the relationship between state and non-state actors in the Carterets/Bougainville case, and it stresses the importance of a 'bridging institution' like Tulele Peisa (an NGO in the civil society context, but grounded in traditional customary structures). Such bridging institutions can build connections between the customary life worlds of the people in the villages and their traditional knowledge on the one hand and the 'outside' world of state, international organisations, donors, NGOs and other civil society organisation and their climate change parlance on the other.

Such bridging institutions can also address an important shortcoming of the current mainstream climate change adaptation discourse: "sidelining God" (Nunn 2017). To be more precise, research which understands the climate change–conflict nexus and conflict-sensitive climate change governance and adaptation in merely secular terms, and climate change adaptation programmes and projects which are designed and implemented as secular endeavours, miss a crucial dimension of the problematique and are bound to fail. John Campbell in his chapter cautions against climate change response strategies and projects dominated by international agencies informed by 'western' scientific, engineering and economic practices as they miss the spiritual and emotional dimensions of the problems at hand. In a similar vein, Kate Higgins in her chapter draws attention to the role of the churches in climate change adaptation and criticises externally led secular projects which solely 'explain science' to communities and impose foreign secular language and approaches upon them; she explicitly points to conflict risks associated with externally designed projects in climate change adaptation and disaster risk reduction. This resonates with Patrick Nunn's research finding

> that one reason for the failure of external interventions for climate-change adaptation in Pacific Island communities is the wholly secular nature of their messages. Among spiritually engaged communities, these secular messages can be met with indifference or even hostility if they clash with the community's spiritual agenda.
>
> (Nunn 2017)

Accordingly, Nunn warns against "sidelining God" in climate projects. And Upolu Luma Vaai cautions that "climate solutions from a secular perspective (…) may not touch deeply the unseen wounds of societies"; he emphasises that "the search for climate change solutions must go beyond the secular" (Vaai 2019, 4).

These reflections on the shortcomings of secular approaches lead to my fourth and final point: the need for a fundamental epistemological (and methodological) shift in research and practice. So far, different non-Western cosmologies, ontologies and epistemologies have not been given much attention in the international discourse on climate change and its effects, including conflict. Pacific experiences and perspectives can change this. They may even provide avenues for the development and implementation of climate change programmes that can support peacebuilding in innovative ways, in particular when drawing on the concept of (eco-)relationality as understood in Oceania.

The concept of (eco-)relationality embraces "both individuality and communality, unity and diversity, visibility and invisibility, male and female, top and bottom, secular and sacred, heaven and earth, God and the world, (…), tangible and intangible. Relationality is a both/and way of thinking" (Vaai and Nabobo-Baba 2017, 11; see also Vaai 2019). Being relational "is about wrestling to understand the 'individual' as part of the 'community' and the 'community' as imaged in the 'individual' (…) it is about being able to have a fluid and holistic grasp of both" (Vaai 2017, 26). Affects, emotions, feelings are integral to such a relational understanding of self, community and, consequently, peace.

Such a relational ontology gives priority to relations over entities.[3] It takes human beings not as isolated 'individuals', but as members of communities, defined through their—not only rational, but also affective and spiritual—relationships with other human beings as well as with actors beyond the human sphere, in nature and the spirit world.[4] Accordingly, peace(building) as well as climate change adaptation in a relational Pacific understanding cannot be had without the inclusion of the spirit world.

Personhood in Pacific communities is (understood as being) genuinely "relational and contextual" (Nabobo-Baba 2017, 163; see also Nanau 2017). And community is not understood in an anthropocentric way, but in a holistic cosmic way, including people, land, ocean, ancestors, spirits, trees, villages, animals, language, mountains, God—who all exist only relationally. Consequently, the 'environment' or the 'climate' cannot be understood in an anthropocentric, dualistic and substantialist manner (as separate from people, society and the sacred), but have to be understood cosmologically. Accordingly, "we can never confine the climate change discussions to the physical material dimensions; rather, we have to take into serious consideration the spiritual dimension that shapes the being and structure of the multiple relationships in the household" (Vaai 2019, 13)—with household understood in the encompassing way inherent in the Greek 'oikos' as an inclusive, holistic, relational concept embracing the whole inhabited earth (Bhagwan 2019; Vaai 2019).

This has far-reaching consequences for (the study of) climate change adaptation and governance, conflict prevention and peacebuilding. The dominant narrow Western approach has to be overcome and Western ways of knowing have to be aligned and combined with traditional Indigenous knowledge—from the Pacific and elsewhere. Cultural sensitivity and cross-cultural dialogue are indispensable for dealing with climate change, conflict, peace and security in research, policy and practice.

More than a decade ago, Dan Smith and Janani Vivekananda had already pointed to the dangers of cultural insensitivity:

> To ordinary people it will feel like outside experts coming and telling them how things are, how they should live and what they should do. The likelihood is that they will ignore this advice or, if necessary, fight it. A different way of working is possible, grounded in a peace-building approach. This emphasises the importance of local knowledge and seeks the active participation of local communities in working out how best to adapt to climate change.
>
> (Smith and Vivekananda 2007, 29).

Accordingly, they identified the need "to bring hard science and local knowledge together" (ibid.), acknowledging "that local knowledge alone is not enough, because climate change throws up unprecedented problems, but nor is the best hard science enough by itself, because adaptation needs to be locally grounded and culturally appropriate" (Smith and Vivekananda 2007, 32). A decade later, Western researchers repeat the same plea: "Appreciating local agency and perspectives also allows for including indigenous and informal knowledge into the assessment of risks and the development of strategies to enhance resilience" (Schilling et al. 2017, 114).

The IPCC also supports the incorporation of Indigenous knowledge into adaptation planning in small island states (Nurse/IPCC 2014, 1636) and criticises that "such forms of knowledge are often neglected in policy and research" (Adger/IPCC 2014, 2). It holds that "mutual integration and co-production of local and traditional and scientific knowledge increase adaptive capacity and reduce vulnerability" (Adger/IPCC 2014, 10). Taking local knowledge seriously also means giving due consideration "to locally appropriate means for knowledge transmission" (McCarter et al. 2014, 8). This necessitates bringing in the above-mentioned 'bridging' institutions and actors who are familiar with both worlds, the local world of traditional knowledge and the international world of scientific knowledge. In this context, however, one has to be cautious not to expropriate traditional, indigenous, local knowledge. Morgan Brigg reminds us: "The inclusion of Indigenous voices and perspectives frequently leads them to be processed in ways that meets the conventions of Western scholarship" (Brigg 2016a, 156). He points to the tendency to just incorporate Indigenous knowledge into Western knowledge and formats according to the latter's terms and thus again subordinate Indigenous knowledge. He therefore recommends to proceed "with extreme caution (....) because the phenomenon of the white man claiming access to Indigenous forces is a longstanding trope in the effacement and replacement of Indigenous people by colonisers" (ibid.). Indeed, Pacific Islanders have hurtful experiences of their wisdom and stories stolen by outsiders, or being forced to express their experiences and knowledge in alien Western formats. Accordingly, they have the suspicion that acknowledgement of 'traditional knowledge' by outsiders will only lead to another wave of quasi-colonial exploitation of that knowledge and to more epistemological violence. Genuine cross-cultural dialogue has to be different; it needs deep self-reflection and immense effort to challenge one's own deeply ingrained convictions about the 'right' way to see the world and how to change it for the better.

Openness to genuine cross-cultural dialogue not least means to engage with indigenous cosmologies, ontologies and epistemologies and to aim at co-production of knowledge together with the 'subjects' of research—what recently was termed ethnographic peace research (Millar 2018): to conduct research in close collaboration with local researchers and affected communities whose voices still are largely absent from the climate change and conflict discourse today. 'Weaving the mat' is an excellent Pacific metaphor for such an approach. Pacific approaches might introduce relational-affective non-anthropocentric perspectives that can provide insights and entry points for policy and practice which so far have been missed by the Western-dominated international discourse.

Fine-grained ethnographic peace research which pays attention to the complexity of local context, the micro, the everyday (Millar 2018, 267) can not only fill current knowledge gaps, but it can also provide urgently sought-after recommendations for policy and practice. It advances "understanding of inter-linkages between governance and the cultural and social context, which is important for a thorough assessment of local adaptive capacity and resilience" (Petzold and Ratter 2015, 42), not least with regard to conflict prevention and peacebuilding, and on this basis can give concrete policy and strategic guidance to policymakers and practitioners. It allows for putting forward highly localised and specific policy recommendations tailored to particular cultural–social–political contexts. Such research thus can take on "the urgent challenge to move from analysis to action on addressing climate-fragility risks" (Vivekananda 2017, 2), addressing practitioners' and policymakers' need for more input from the academic realm, so that they can develop well-informed and targeted policies, strategies, governance and adaptation measures.[5]

Policy-relevant research along these lines will not only have to explore the conflict-prone effects of climate change, but also the conflict potential of climate change adaptation and mitigation policies and technologies, as Kate Higgins reminds us in her chapter. And it can even go a decisive step further by exploring the "climate-cooperation nexus" (Ide et al. 2016, 297), that is: exploring the potential of climate change policies for building and sustaining peace—as Kirsten Davies has started to do in her chapter, pointing to climate change as a community connector. Accordingly, Halvard Buhaug in his chapter reminds us not to ignore success stories. The chances of success are multiplied by linking climate change policies and peacebuilding policies, with a climate-sensitive approach to peacebuilding, and a conflict-sensitive approach to climate change adaptation (Nett and Ruettinger 2016, 49, 56; Taenzler, Ruettinger and Scherer 2018, 4). If conflict-sensitive climate change adaptation and climate-sensitive peacebuilding are woven together, the possibility arises to even use climate change induced disasters and crises "as opportunities to build peace and increase the resilience of affected populations" (Nett and Ruettinger 2016, 53).

The contributions in this book show that Pacific-focused research can make an important contribution to the wider international debate about climate change, conflict, peace and security. They widen the perspective beyond conventional mainstream approaches, taking into account so far underestimated or marginalised aspects. Hence this book makes an important start for putting the Pacific on the map of the international climate change, conflict, peace and security discourse—and for putting researchers from the Global North and from the Pacific region into dialogue. The editors of this volume are planning to continue this undertaking in the future with a book series on Climate Change and Conflict in Oceania. This series is supposed to contribute to dialogue across difference, between different cultures, different ways of knowing, different epistemologies, different academic traditions: building bridges, weaving mats.

Notes

1 On the more recent debate about ontological security in peace and conflict studies, see, for example, Kinnvall (2017), Jarvis (2018), Kinnvall and Mitzen (2018) and Croft and Vaughan-Williams (2017). Farbotko (2019) links climate change displacement and ontological security in a Pacific context and flags exploration of this link "as a priority research need" (Farbotko 2019, 255).
2 "Quantitative large-N studies are currently the most widely accepted methodological approach in the research on climate change and violent conflict, although they face severe problems regarding the quality of their data sets and their ability to capture complex human-nature interactions" (Ide et al. 2016, 288).
3 While relationalism gives ontological precedence to relations, interactions and flows, substantialism in contrast prioritises entities, units and structures that are bound and fixed (Brigg 2016b, Hunt 2017, Hunt 2018).
4 On relationality in a Melanesian context and the understanding of the Melanesian 'va'—'the space between', that is: a relational space that separates and joins, not least connecting the spiritual and the secular—see the contributions in Vaai and Nabobo-Baba (2017).
5 In this context it is worth noting that the case studies in this book speak to the research desiderata identified by the IPCC in regard to small island states. The IPCC contends that there is need to acknowledge the "heterogeneity and complexity of small island states and territories" (Nurse/IPCC 2014, 1644). Accordingly, "within-country and—territory differences need to be better understood" (ibid.); in particular there is "need for more work on rural areas, outer islands, and secondary communities" (ibid.). This book's case studies have made a start in addressing this need, and it is planned to pursue this avenue further.

References

Adger W. Neil, Juan M. Pulhin, Jon Barnett, Geoffrey D. Dabelko, Grete K. Hovelsrud, Marc Levy, Ürsula Oswald Spring, et al. IPCC 2014. "Human Security". In *Climate Change 2014. Impacts, Adaptation, and Vulnerability. Part A: Global and Sectoral Aspects*. Contribution of Working Group II to the Fifth Assessment Report of the Intergovernmental Panel on Climate Change, edited by C.B. Field, V.R. Barros,

D.J. Dokken, K.J. Mach, M.D. Mastrandrea, T.E. Bilir, M. Chatterjee, K.L. Ebi, Y.O. Estrada, R.C. Genova, B. Girma, E.S. Kissel, A.N. Levy, S. MacCracken, P.R. Mastrandrea, and L.L. White, 755–791. Cambridge: Cambridge University Press.

Bhagwan, James. 2019. *Climate Initiatives Must Change*. Pacific Conference of Churches. https://pacificconferenceofchurches.org/f/climate-initiatives-must-change (accessed 20 January 2020).

Boe Declaration. 2018. *Forty-Ninth Pacific Islands Forum, Boe Declaration on Regional Security*. 6 September 2018. https://www.forumsec.org/boe-declaration-on-regional-security/ (accessed 20 January 2020).

Brigg, Morgan. 2016a. "Engaging Indigenous Knowledges: From Sovereign to Relational Knowers." *The Australian Journal of Indigenous Education* 45 (2): 152–158.

Brigg, Morgan. 2016b. "Relational Peacebuilding: Promise beyond Crisis." In *Peacebuilding in Crisis: Rethinking Paradigms and Practices of Transnational Cooperation*, edited by Tobias Debiel, Thomas Held, and Ulrich Schneckener, 56–69. London and New York: Routledge.

Bryant-Tokalau, Jenny. 2018. *Indigenous Pacific Approaches to Climate Change: Pacific Island Countries*. Cham, Switzerland: Palgrave Macmillan.

Brzoska, Michael, and Christiane Fröhlich. 2015. "Climate Change, Migration and Violent Conflict: Vulnerabilities, Pathways and Adaptation Strategies." *Migration and Development* 5 (2): 190–210.

Croft, Stuart, and Nick Vaughan-Williams. 2017. "Fit for Purpose? Fitting Ontological Security Studies 'Into' the Discipline of International Relations: Towards a Vernacular Turn." *Cooperation and Conflict* 52 (1): 12–30.

Farbotko, Carol. 2019. "Climate Change Displacement: Towards Ontological Security." In *Dealing with Climate Change on Small Islands: Towards Effective and Sustainable Adaptation?*, edited by Carola Kloeck, and Michael Fink, 251–266. Goettingen: Goettingen University Press.

Hunt, Charles T. 2017. "Beyond the Binaries: Towards a Relational Approach to Peacebuilding." *Global Change, Peace & Security* 29 (3): 209–227.

Hunt, Charles T. 2018. "Relational Perspectives on Peace Formation: Symbiosis and the Provision of Security and Justice." In *Exploring Peace Formation: Security and Justice in Post-Colonial States*, edited by Kwesi Aning et al., 78–99. London and New York: Routledge.

Ide, Tobias, Peter Michael Link, Jürgen Scheffran, and Janpeter Schilling. 2016. "The Climate-Conflict Nexus: Pathways, Regional Links, and Case Studies." In *Handbook on Sustainability, Transition and Sustainable Peace*, edited by Hans Günter Brauch et al., 285–304. Springer.

Jarvis, Lee. 2018. "Toward a Vernacular Security Studies: Origins, Interlocutors, Contributions, and Challenges." *International Studies Review* 2018 (0): 1–20.

Kinnvall, Catarina. 2017. "Feeling Ontologically (In)secure: States, Traumas and the Governing of Gendered Space." *Cooperation and Conflict* 52 (1): 90–108.

Kinnvall, Catarina, and Jennifer Mitzen. 2018. "Ontological Security and Conflict: The Dynamics of Crisis and the Constitution of Community." *Journal of International Relations and Development* 21 (4): 825–835.

Klare, Michael T. 2019. *All Hell Breaking Loose: The Pentagon's Perspective on Climate Change*. New York: Metropolitan Books.

McCarter, Joe, Michael C. Gavin, Sue Baereleo, and Mark Love. 2014. "The Challenges of Maintaining Indigenous Ecological Knowledge." *Ecology and Society* 19 (3): 39.

Millar, Gearoid 2018. "Decentring the Intervention Experts: Ethnographic Peace Research and Policy Engagement." *Cooperation and Conflict* 53 (2): 259–276.

Nabobo-Baba, Unaisi. 2017. "In the Vanua: Personhood and Death within a Fijian Relational Ontology." In *The Relational Self: Decolonising Personhood in the Pacific*, edited by Upolu Luma Vaai, and Unaisi Nabobo-Baba, 163–175. Suva: University of the South Pacific Press.

Nanau, Gordon Leua. 2017. 'Na Vanuagu': Epistemology and Personhood in Tathimboko, Guadalcanal. In *The Relational Self: Decolonising Personhood in the Pacific*, edited by Upolu Luma Vaai, and Unaisi Nabobo-Baba, 177–201. Suva: University of the South Pacific Press.

Nett, Katharina, and Ruettinger, Lukas. 2016. *Insurgency, Terrorism and Organised Crime in a Warming Climate: Analysing the Links between Climate Change and Non-State Armed Groups*. Berlin: Adelphi.

Nuccitelli, Dana. 2019. *Climate Change Poses Security Risks, According to Decades of Intelligence Reports: The Long History of Climate Change Security Risks*. Yale Climate Connections. https://www.yaleclimateconnections.org/2019/04/the-long-history-of-climate-change-security-risks/ (accessed 20 January 2020).

Nunn, Patrick D. 2017. "Sidelining God: Why Secular Climate Projects in the Pacific Islands are Failing." *The Conversation*. May 17, 2017. http://theconversation.com/sidelining-god-why-secular-climate-projects-in-the-pacific-islands-are-failing-77623 (accessed 20 January 2020).

Nurse, L., R. McLean, J. Agard, L. P. Briguglio, V. Duvat-Magnan, N. Pelesikoti, E. Tompkins, et al. IPCC 2014. "Small Islands". In *Climate change 2014. Impacts, Adaptation, and Vulnerability. Part B: Regional Aspects*. Contribution of Working Group II Contribution to the Fifth Assessment Report of the Intergovernmental Panel on Climate Change, edited by V.R. Barros, C.B. Field, D.J. Dokken, M.D. Mastrandrea, K.J. Mach, T.E. Bilir, M. Chatterjee, K.L. Ebi, Y.O. Estrada, R.C. Genova, B. Girma, E.S. Kissel, A.N. Levy, S. MacCracken, P.R. Mastrandrea, and L.L. White, 1613–1654. Cambridge: Cambridge University Press.

Pacific Islands Forum. 2019. Fiftieth Pacific Islands Forum Communique. Kainaki II Declaration for Urgent Climate Action Now. https://www.forumsec.org/wp-content/uploads/2019/08/50th-Pacific-Islands-Forum-Communique.pdf (accessed 20 January 2020).

Petzold, Jan, and Beate M.W. Ratter. 2015. "Climate Change Adaptation under a Social Capital Approach – An Analytical Framework for Small Islands." *Ocean & Coastal Management* 112: 36–43.

Rüttinger, Lukas, Dan Smith, Gerald Stang, Dennis Tänzler, and Janani Vivekananda. 2015. "A New Climate for Peace: Taking Action on Climate and Fragility Risks. Executive Summary." An Independent Report Commissioned by the G7 Members. Berlin: Adelphi.

Schilling, Janpeter, Sarah Louise Nash, Tobias Ide, Jürgen Scheffran, Rebecca Froese, and Pina von Prondzinski. 2017. "Resilience and Environmental Security: Towards Joint Application in Peacebuilding." *Global Change, Peace & Security* 29 (2): 107–127.

Smith, Dan, and Janani Vivekananda. 2007. *A Climate of Conflict: The Links between Climate change, Peace and War*. London: International Alert.

Taenzler, Dennis, Lukas Ruettinger, and Nikolas Scherer. 2018. *Building Resilience by Linking Climate Change Adaptation, Peacebuilding and Conflict Prevention*. Planetary Security Initiative Policy Brief. Berlin: Adelphi.

Toda Declaration. 2019. "Toda Pacific Declaration on Climate Change, Conflict and Peace." Toda Peace Institute Policy Brief No.41. Tokyo: Toda Peace Institute. https://toda.org/pacific-declaration.html (accessed 20 January 2020).

Vaai, Upolu Luma. 2017. "Relational Hermeneutics. A Return to the Relationality of the Pacific Itulagi as a Lens for Understanding and Interpreting Life." In *Relational Hermeneutics,* edited by Upolu Luma Vaai, and Aisake Casimira, 17–41. Suva: University of the South Pacific Press.

Vaai, Upolu Luma. 2019. "'We are Therefore We Live'": Pacific Eco-Relational Spirituality and Changing the Climate Change Story." Toda Peace Institute Policy Brief No. 56. Tokyo: Toda Peace Institute.

Vaai, Upolu Luma, and Unaisi Nabobo-Baba. 2017. "Introduction." In *The Relational Self: Decolonising Personhood in the Pacific,* edited by Upolu Luma Vaai, and Unaisi Nabobo-Baba, 1–21. Suva: University of the South Pacific Press.

Vivekananda, Janani, Shiloh Fetzek, Malin Mobjörk, Amiera Sawas, and Susanne Wolfmaier. 2017. *Action on Climate and Security Risks: Review of Progress 2017.* Den Haag: Clingendael.

Index

For Product Safety Concerns and Information please contact our EU
representative GPSR@taylorandfrancis.com
Taylor & Francis Verlag GmbH, Kaufingerstraße 24, 80331 München, Germany